HOW FAR THE
LIGHT REACHES

ALSO BY SABRINA IMBLER

Dyke (geology)

HOW FAR THE
LIGHT REACHES

A Life in Ten Sea Creatures

SABRINA IMBLER

with illustrations by Simon Ban

Little, Brown and Company
New York   Boston   London

Little, Brown and Company
Hachette Book Group
1290 Avenue of the Americas, New York, NY 10104
littlebrown.com

First Edition: December 2022

Little, Brown and Company is a division of Hachette Book Group, Inc. The Little, Brown name and logo are trademarks of Hachette Book Group, Inc.

The Hachette Speakers Bureau provides a wide range of authors for speaking events. To find out more, go to hachettespeakersbureau.com or call (866) 376-6591.

ISBN 9780316540537
LCCN 2022936696

10 9 8 7 6 5 4 3 2 1

MRQ-T

Printed in Canada

What does the light want?
More of its kind?
Yes.
Yes and
a wish to disturb the dark.
　—from Kimiko Hahn, *Resplendent Slug*

Contents

Contents

If You Flush a Goldfish

The truth is that I was asked to leave the Petco, but I told everyone I was banned. The word carried more weight, more daring, more drama than my thirteen-year-old life had ever seen. I was asked to leave just one particular Petco, the one in the shopping center built on a landfill by my hometown, but I told everyone I was banned from Petco, implying Big Petco, hoping they might assume the entire chain had deemed me a threat to their business.

I had come to Petco to stage a protest in the aquarium section. My demonstration went like this: I stood by the fishbowls and tried to convince the occasional customer not to buy them. The Petco I had chosen — the one closest to my house — was mostly empty, and so my demonstration could have passed for me quietly shopping. The real shoppers, very few of whom had come to buy goldfish bowls, it seemed, ignored me. Occasionally, when someone mistook me for a Petco employee, I stammered an apology and ducked into the reptile aisle. If the aisles were empty, I watched the goldfish tank. It was almost as big as a bathtub, and the orange fish inside it shimmered like sequins. The tank seemed more fish than water, a stampede of glinting scales moving in every direction, searching, perhaps, for some space. The dead and dying fish drifted to the edges of the tank—bloated and bobbing at the surface, lolling half-eaten at the bottom, bent and half-sucked into the filter.

Time passed quietly until a mother approached the shelves I stood vigil by and picked out a glass bowl, presumably for her son, who had wandered away. My carefully practiced argument (keeping a goldfish in a bowl was inhumane) devolved into sporadically recalled facts— Goldfish pee themselves to death in bowls! Goldfish can grow up to a foot long! Goldfish can live up to twenty years!—until a blue-poloed Petco sales associate told me I needed to leave. I had to call my mom to come and pick me up in the parking lot, where a different Petco sales associate waited with me until her beige SUV appeared on the horizon.

We—the Petco sales associate and I—were just a mile away from San Francisco Bay. It was the closest I usually got to something like a sea, and if I closed my eyes I could taste the salt in the air. When the wind died and the pungent ocean smell wafted away, I could nose out another, heavier odor: trash, so faint you would second-guess your-self if it didn't keep coming back, the distinct funk of some-thing, somewhere rotting.

As we waited, smelling salt and trash, my incompetence nauseated me. My first attempt to help something I cared about, and I'd failed. All those doomed and dying fish. The luckiest of them might go to an aquarium. The rest would end up dead in bowls, though they wouldn't die right away. It is nearly impossible to hurt yourself while living in a fish's equivalent of a padded cell: smooth, uncornered glass that could never even scrape a scale. But every single one would die eventually, probably before its time. It would die because someone had forgotten to care for it, or decided that caring for it properly was too much work. Too much work to empty its dirty water and replace it with fresh. Too inconvenient to provide it with enough space to live and grow.

At the time, the best future I could imagine for the gold-fish was life in a bigger tank, maybe even thirty gallons, with fresh water and some plastic plants. A more comfort-able confinement. Having only ever seen goldfish crowded in Petco tanks or isolated in bowls, I had no conception of what their lives might look like outside the glass walls of an aquarium. I could not imagine what a goldfish is capa-ble of becoming in the wild.

Back then I assumed the shopping center by the Petco smelled like trash because the shopping center was built on a landfill. My mom had told me the whole city was built on a landfill, and I envisioned buildings perched on condensed slabs of trash. But the land beneath the Petco was once salt marsh in a vast expanse of wetlands wrapping around San Francisco Bay. Today, satellite images of the bay show a sharp divide between green and blue, but hundreds of years ago there was no clear division between land and sea. The bay was an estuary, salt and fresh water intermingling into brackish water. Lapping waves and shifting tides exposed and swallowed land each day. At lower elevations, the claggy, salty soil was (and remains) inhospitable to most plant life. But higher up, native plants thrived: Pacific cordgrass grew as tall as teenagers, interspersed with beds of rosy-fingered pickleweed. This was the nature of the bay for ten thousand years when Indigenous peoples, including the Coast Miwok, or the many groups of the Ohlone, such as the Muwekma, Ramaytush, Tamien, Chochenyo, and Karkin peoples, lived there and foraged in the marsh.

The Spanish arrived in the 1700s and baptized, enslaved, and indirectly massacred the Ohlone people with disease. Starting about 150 years ago, more recent settlers had ambitions of developing the bay into farms and towns, but you cannot grow crops or build a home on a salt marsh. So the wetlands were seen as useless and disposable, and destroyed. The bay was diked and the waterlogged soil desiccated into silty mud. The land became a dairy farm, with cows and hayfields and salt ponds. In the 1960s, the land was zoned for single-family housing, and millions of cubic yards of sand and mud were dumped on the former tidelands so that buildings would not sink through the soft silt and into the ocean. The land was called reclaimed marshland, and the streets carved into the ground were named after the wild things that had been forced out: Oyster Court. Pompano Circle. Flying Fish Lane. As a child, I did not know there were two meanings to "landfill." I did not know that the reek of the Foster City Petco parking lot may have come from the bay itself, water fouled by multiple petroleum refineries, the wastewater treatment plants, the coughing black pipes of ships.

By the time I was born, San Francisco Bay had lost 95 percent of the wetlands and salt marshes that once had collared the sea. All two hundred thousand acres of tidal channels, mudflats, sandbars, streams, and pools conjured only during floods had been paved over into farms and cities and factories and military bases and tourist towns and freeways and a Petco. This is to say: I knew my hometown as a suburb and never imagined what it had been before. I couldn't wait to leave.

If I could do it over again, this is what I would tell the mother in that Petco:

You may have read that a goldfish grows in proportion

to the size of its bowl. But unlike us, goldfish are indeterminate growers; if given the chance, they will grow until they die. Different kinds of goldfish can grow into a range of sizes and shapes. In the wild, an adult goldfish can weigh as much as a pineapple.

You may think goldfish live for just a year, maybe two. But they can actually live much longer. Twenty years, if they're lucky. Goldfish can survive a few years in a bowl because they are almost supernaturally hardy, capable of weathering conditions that would quickly kill most other fish. A bowl is a tiny, isolated environment starved of oxygen, which means even a slight change in the water chemistry can be lethal. I say this because goldfish pee with abandon. They unleash more ammonia than other aquarium fish, a toxin that would be diluted in a pond or a river but can kill a fish in a bowl. This is why, I would say to the woman, a bowl makes the conditions of living impossible. But when a goldfish manages to survive it, no one thinks of their feat as extraordinary.

Lastly, I would tell her, you may have heard that goldfish have a three-second memory. But goldfish can remember that a colored paddle means food is coming, even months after the association is formed. Goldfish can perform complex tasks, such as escaping a net or navigating a maze. How can such a small fish hold on to the memory of the snaking path of a maze for three months? Could you do that? What is it like for a creature with a three-month memory to live and die in a bubble the size of a dutch oven?

Whenever I start an internship or a new job, I tell people that I was kicked out of Petco as a teenager. It has become something of an origin myth, my designated fun fact. I have told the story so many times that the details of my original memory have become inaccessible, transformed from a real experience into a rote narrative. I don't remember what I told my mom to get her to drive me there, or how I scrounged up the courage to be antagonistic with strangers when I could barely stand up to my middle-school bullies, whose bland, derivative cruelty still managed to make me hate myself.

I do remember that I was in eighth grade. I remember that I was thirteen—an awful year. I remember that I went to a private school where the door above the headmaster's office was inscribed with a phrase in Latin that translated to "Leisure without learning is death." The first time I met my classmates, a gaggle of kids showed up in sweatshirts emblazoned with STANFORD. I also showed up in a hoodie, mine proclaiming GAP. We were ten. I overheard one student's mom say to another at the mixer, "You know, this is a feeder school for Stanford," and the other mom nodded in agreement. I had never heard the term "feeder" applied to a school before, only to tanks of goldfish and guppies—fish inexpensive and unremarkable enough that aquarists buy them as live prey for their larger, more valuable pets.

I remember that many of my classmates were the children of powerful people: board members and professors at Stanford, executives from Silicon Valley and Morgan Stanley, heiresses. These children had last names like Packard and Jobs. The orientation pool party took place at one of their houses, which appeared to me like a castle, with two pools and, across a fountained emerald lawn, a tennis court. I know that my parents sent me to this school in part so that I could get into the best possible college, which they believed meant I would live the best possible life. I reminded myself of this as the heir to a computer technology company chased me in loops around the padded walls of the gym during Friday's free period, wielding a plastic segmented jump rope like a whip.

I lived a few blocks from that school, and I remember rich kids would drive down my street as if they were impervious to death. I would hear the telltale screech of tires swerving and jump into the closest driveway or hedge and watch the cars barrel past. I remember watching a metallic luxury SUV turn out of the school's driveway and skid into our mailbox. The car hurtled on, leaving behind a white metal frame wrenched like an elbow, red flag dangling like a broken arm. I remember that kids at schools near mine were killing themselves because of all the pressure, enough suicides for the CDC to deem the deaths a "cluster." I remember that one student's obituary included his ACT scores. Another student's obituary listed her number of Facebook friends. I remember I spent late nights on AIM trying to talk my friend out of wanting to die.

I had terrible insomnia back then, and I remember lying awake at night, trying to imagine the best possible version of my future, which always assumed a similar form. After college, a vaguely important job where I wore blazers and pencil skirts. A husband (ideally hot) after a respectable number of boyfriends. Finally, clear skin. But when I tried to fantasize about these rote and sensible futures, my mind always wandered to my death. Specifically, I imagined my funeral—what it would look like, who would attend, who I would have my funeral bouncer turn away at the door (I had clearly never been to a funeral before). It wasn't exactly that I wanted to die but that ceasing to exist (and being reverently mourned) felt more tangible to me than what I had been told I should want.

I got my first and only goldfish from that middle school. It was part of a science project, and our biology teacher, who always smelled of hemp, announced that anyone who wished could take a goldfish home with them. She did not tell us what would happen to the fish if we did not take them home, and we did not think to ask. I named the fish Quincy and kept it in a bowl on my dresser. Sometimes

Quincy swam, but mostly it floated. Its body seemed suspended by string, fins twitching without purpose around the chintzy castle and honeydew-colored kelp I had rooted in the marbles at the bottom of its bowl. I spent a lot of time watching Quincy. When I considered, even briefly, how little space the fish had to move and grow, I wondered if I was doing something cruel.

So I asked my dad to drive me to the Japanese garden in our local park. I smuggled Quincy in a small jar in the cavernous pocket of my Gap sweatshirt, walked to a shrouded corner of the koi pond, and overturned the jar. Quincy's orange body wriggled into the murk, and then, at last: relief.

When I visited the garden months later, I looked for Quincy but never found it.

Sometimes, when people learn they are killing their goldfish, or when they have grown bored of their pets, they dump them. Sometimes they dump them in Japanese garden ponds. More often they dump them in larger water bodies: lakes, creeks, rivers. If goldfish are doomed in a bowl, they are unstoppable in a river. They do more than survive; they take over the whole place. Their gills, once rouged by the ammonic burn of their piss, drink in the oxygen of surging, aerated water. Gorged on algae and worms and snails and the eggs of other fish, their bodies begin to balloon. They swell to the size of Cornish game hens, cantaloupes, jugs of milk.

These are feral goldfish, and if you saw one you might not recognize it. Goldfish that are actually gold will revert to their natural colors in a matter of generations. Bright orange fish disappear, eaten by predators, succeeded by fish with duller sheens. They become somewhat indistinguishable from other carp. They disappear within the weeds.

Feral goldfish are so good at living they have become an ecological menace. Of course, it's not their fault; goldfish would never have gotten into the river if we hadn't thought of them as disposable. Wild goldfish have been found in every state but Alaska, and when they are unleashed in a water body, they ruin whatever balance life might have found before. Their riotous presence drives out native species. Goldfish love to dig, and they will uproot everything growing at the bottom of a lake in search of something to eat. When they devour opaque clouds of cyanobacteria, their intestines foment the bacteria's growth, making them incubators of algal blooms. They may spawn as early as a year old, releasing hundreds of sticky eggs that cling to rocks and plants and anything that will hold them.

Once goldfish are in a pond or lake or river, you cannot remove them. You cannot reel or net them all out, and however many goldfish you have taken will be replenished when they breed again. The only way to kill the goldfish is to kill every fish along with them in the water, dumping gallons of rotenone, a biocide poisonous to fish, to ensure nothing can survive. But even this is only possible in ponds and lakes, water bodies with hard edges where the poison will not escape.

One river in southwestern Australia has become overrun with feral goldfish, all descended from a handful of pets someone dumped two decades ago. The balmy conditions of this river, called the Vasse, are goldfish paradise, and here they grow faster than any other population of ferals. Most of the goldfish in the Vasse are the colors of earth—browns and olives and dark greens—but some of the largest ones are unmistakably orange. These bigwigs, weighing as much as a butternut squash apiece, are likely the original goldfish dumped in the Vasse, or their direct offspring. Do these goldfish remember, even faintly, what life in a bowl was like?

One scientist who tracks the ferals in the Vasse has noticed they are capable of remarkable things. He saw schools of goldfish traveling nearly 1,000 feet per day. One fish covered more than 140 miles in one year. The entire community of feral goldfish migrated seasonally, swimming in vast shoals to a distant wetland during the breeding season. The goldfish, raised in captivity or born in a river they were never supposed to be in, possessed a seemingly innate knowledge, preserved across generations of bowled fish.

Scientists have also discovered feral goldfish in estuaries. At first, they assumed goldfish could not penetrate wetlands where fresh water mingles with salt. But the more scientists look, the more they find goldfish in waters nearer and nearer the sea. One population sprouting from the Vasse River seemed to have developed a higher tolerance to salt than any other goldfish population in the world. Scientists wondered if this population was a possible sign — that salt-resilient goldfish could use estuaries as salt bridges to migrate to new rivers, new lakes. The feral goldfish of the Vasse have unknowingly come closer to the ocean than any other goldfish we know of. They have confronted seemingly inhospitable waters, and they have survived.

Maybe there is something universal about wanting to get out. I wonder if the goldfish have any sense of the ocean that lies ahead of them.

When my parents decided I should transfer to another high school, I wept and raged. I offered to pay them back for my tuition by selling my blood plasma, by participating in medical experiments, by selling an egg. "Don't be ridiculous," my mom told me, horrified. "You're not even old enough to sell an egg."

At my new school, I overcompensated to salvage my

chance at the future I had been told I should want. I took extra classes, extra extracurriculars. Anything for Stanford. I went to school at 7 a.m. for biology and stayed until after 11 p.m. for newspaper. On weekends, I volunteered at football games, spiraling slugs of golden cheese on nachos and constructing flat, horrible hamburgers. I filled every waking moment with a task; no one could say I hadn't tried hard enough, worked hard enough, given everything I had to give. At least now I slept easily, short and dreamless nights that ended in a stack of alarms. I could no longer sense who I was, what "happy" might look like, because there was always something I had to think about. Meaning: I was insufferable and you probably would have hated me. I hated me.

The summer before I left for college, I volunteered on a research boat in San Francisco Bay. I had gotten my license and relished the drive over, windows of my mom's beige SUV rolled down to let in the salty air. I spent four-hour shifts on the boat, a 90-foot-long vessel with a hull the color of deep water. The captain looped us around the estuary so we could siphon samples of mud and water. During the voyage, we would fling an otter trawl net over the back of the boat, let it drag for ten minutes, and haul it back in, spilling the contents of the net into creamy white tanks that sprouted from the deck like toadstools.

My job was to measure and identify every creature we caught. My first days on the boat, I was useless, squinting into the sea spray, my arms scorched red despite the blanket of fog that rolled over the bay. Everything on deck was slippery, and I kept dropping my clipboard. One by one, I scooped the fish up out of the tank and laid them flat against a clear ruler glued to a table by the tank. I smoothed out their thrashing, mucus-slicked bodies, talking to them as if I could coax them to lie still, which I never could. In my hands, the fish moved in a way I'd never seen bodies move before: seconds of impeccable stillness, the only moving thing an eye, flickering wildly, and then the entire body would arch as the fish flung itself up into the air. Fish leaping, pirouetting, somersaulting into sky. And me, scrambling after their landed bodies, cupping them in my hands until I could tip over the railing and plunge them back into water.

Some days we caught only anchovies and sardines, and my head spun distinguishing between five hundred, six hundred, seven hundred nearly identical creatures. But occasionally the nets would deliver us something wonderful. A speckled tonguefish, its two swiveling eyes looking directly at me as I measured it. California bat rays, wings flapping toward the rims of the tank as if they knew what flight was and wanted to try it. Once we caught two baby leopard sharks and I learned how to hold one, my left hand wrapped around the tail and my right cupping the spot below its small, jagged mouth. The shark wriggled, snake-like, but I held it firm and fast until it was time to let it go.

Because the bay is an estuary, we caught fish that could live in brackish water. Starry flounders, chameleon gobies, staghorn sculpins that prickled their spines and jabbed you if you held them too tight. Someone once caught a white sturgeon, its skin the color of a pearl and its body more than a hundred pounds.

We caught non-native species too, of course. San Francisco Bay is often called a "highly invaded ecosystem," one of the most invaded estuaries in the world. In certain habitats, these introduced species outnumber the native ones and outmass them with the sheer weight of their accumulated bodies.

After I measured each fish, it was my job to throw them overboard. At first, I flung them off the side of the boat and turned around before I saw them splash. I imagined the cooling feel of the water against their scales, the rush of oxygen to their gills—disorientation for a moment, and then relief. It took me several days to realize that half of the fish I threw overboard never made it back into the water, instead scooped up by a roving gang of gulls and ospreys that lurked around our boat. The ospreys watched me as I worked, talons gnarled around the white-painted railings, and when I extended my arm overboard, fish in hand, they dove. When the birds crept too near my fish-laden tanks, I waved my slimy clipboard at them. Sometimes I shouted, spit flying, as I raced against the ospreys to pick up a fish that had flung itself out of the water and lay, gasping, on the deck. When I dumped the fish overboard, I leaned over the rails, as close to the surf as I could get without falling, to watch their scales flash and disappear.

On our way back to shore, I plunked my hundreds of anchovy measurements into a computer inside the plastic walls of the boat. And then I sat out on the deck and let the dried salt flake off my skin. I was always covered in scales after trawls. I would hold my arms up to the light and admire my iridescent, sweat-dewed fish-skin. Deep in the bay, when the sky was clear, I felt like I could peer to the very curve of Earth, imagine it dipping. Like I could see into all of my possible futures.

What does it mean to survive in the wild? You can't do it without going wild yourself. We are all capable of reverting to a wilder state. The wild may sentence a cat or a dog to a starved life or early death. But for a goldfish, the wild promises abundance. Release a goldfish, and it will never look back. Nothing fully lives in a bowl; it only learns to survive it.

I will always be a little bit in love with feral goldfish. I know this is the wrong lesson to take from it all. I know they wreak an irreversible kind of havoc. They uproot bottom dwellers, trample ecosystems, sow tasseled parasites in the flesh of other fish. I know that once they take over a pond they are impossible to extricate. I don't want a supremacy of goldfish, a world where fish the size of cantaloupes stampede through fragile ecosystems like wrecking balls. But when I think about ponds infested with gallon-big goldfish, I feel a kind of triumph. I see something that no one expected to live not just alive but impossibly flourishing, and no longer alone. I see a creature whose present existence must have come as a surprise even to itself.

Imagine having the power to become resilient to all that is hostile to us. Confinement, solitude, our own toxic waste. Salt, waves, hundred-pound sturgeon that could swallow us whole. Imagine the freedom of encountering space for the first time and taking it up. Imagine showing up to your high school reunion, seeing everyone who once made you feel small, only now you're a hundred times bigger than you once were. A dumped goldfish has no model for what a different and better life might look like, but it finds it anyway. I want to know what it feels like to be unthinkable too, to invent a future that no one expected of you.

Because the Petco in Foster City is built on a landfill, it is sinking. The landfill is sinking faster than almost anywhere in California, creeping closer to the core of Earth. Each year, the city sinks by as much as 10 millimeters and the sea rises by as much as 3 millimeters. This is a losing battle. For a long time stony levees protected Foster City from the sea. Now water splashes over the levees, onto trails and doorsteps. Soon the waves will overtake the levees and flood the farms, the city, the factories, the military bases, the tourist towns, the freeways, and the Petco. The water may doom the cats, caged birds, leopard geckos, bunnies, hamsters, and guinea pigs. But I like to imagine the fish slipping out of their tanks, heading toward some unknowable horizon.

Years after high school, when I lived in a new and cloudy city where I knew few people, I went back home for a month during the holidays. I had never been back home this long before, and I was surprised by how quickly I fell into old routines. I drove the scuffed beige car to pick up groceries, took my grandparents to the mall, and dropped my sibling off at school. My parents had converted my bedroom into storage, so I slept in a twin bed surrounded by filing cabinets and stacks of CDs. When I jogged in the afternoon, I jumped instinctively out of the way as luxury SUVs barreled down my street out of the gates of my old school.

I lasted a week before I opened Tinder. I told myself I was there to see my old classmates, to see who was newly hot, newly gay, or both. Two people from my Girl Scout troop were on the app, the shy one who was allergic to asparagus and the short one who ate a roly-poly after we dared her to, all of which made sense. I learned someone on my old improv team was trans, and we started following each other on Instagram. And then I saw a familiar face, at first uncanny until I swiped through and realized I had gone to high school with this person. We had never spoken, but I knew exactly who they were. Sometimes, when I walked home from school, I would notice them playing tennis at the school courts. They always wore their hair in a ponytail, white visor sweeping strands out of the way. When they jumped up to serve, my eyes would follow them, turning up toward the sun and becoming momentarily blinded by the brightness of it all. I was never sure at the time why I wanted to keep watching.

We matched, messaged, and I drove to their house, one of the many built on the landfill on the estuary. Their street name had a word like "Sea" in it, and whenever I typed their address into Google Maps, I felt like I might be instructed to drive my car straight into the ocean. I knocked; they let me in. I took off my boots and tucked them carefully by the door, because we both have Chinese mothers, and we padded softly to their room. We spent the next twelve hours sitting on opposite ends of their twin bed. We drank green tea to stay awake. We took turns petting their cat. When I got hungry at midnight, they gave me a Nature Valley granola bar, and when I took a bite, I felt like I was on a field trip. Each time they left to brew more tea, I marveled at how much their room resembled mine: the same school yearbooks, the same edition of *As I Lay Dying* that we had read for AP English. I remembered the scene in which Vardaman calls his dead mother a fish because that was the only way he could understand death.

In the blue light, voices rasped after hours of talking, our heads each propped up with an arm, we looked at each other with an eerie intent. Maybe we were hoping for some sign of what this night meant, both of us afraid of misreading what felt like a dream. Maybe each of us was studying the other's face to see how it had changed, how it had grown. We looked nothing like we did in high school: no more makeup, long hair chopped off and sides buzzed, constellations of tattoos on our arms. We both had been expected to be daughters but turned out to be something else. We had shed our skins, not like snakes but insects — each of us a nymph outgrowing exoskeleton after exoskeleton, and morphing as we did. We didn't know which molt would be our last, only that we might not be there yet, both of us rivers moving toward the sea. A few years after that night, they changed their name and pronouns; even later, I changed mine.

They kissed me at dawn, sun slivering into the room through the blinds. We were both half-asleep after staying up the whole night in anticipation, and as we touched, I felt like I was wafting out of my body. Determined not to disturb the cat still asleep at the foot of the twin bed, we made a wreck of the pillows, our heads crinkling against the slatted blinds. I was so tired I think I started crying, but it was impossible to tell because there was salt water everywhere, crusted on our hands and our faces, dripping from our armpits, our bodies leaking out of themselves. I kept repeating, "I can't believe it," as they grasped my hands. "I can't believe it," as they pressed into my bone. When they asked what I meant, I had no idea what to say. How I couldn't believe we had spent so many years so close, circling the same hallways, and yet had never spoken. How I couldn't believe who either of us had become in the years since, and how each of our becomings felt like an unthinkable triumph. How I couldn't believe how much I wanted to sleep. How I couldn't believe how ridiculous and deeply gay it was that we spent most of a night on a bed petting a cat, unsure if the other person was into it. So I said, "I can't believe it," over and over and they said, "I can't believe it," back. And then I left, and we let each other go.

My Mother and the Starving Octopus

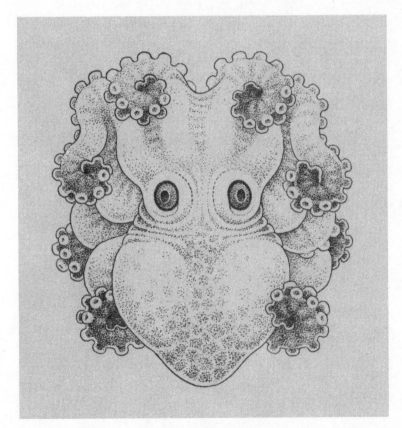

Years ago, when I was in seventh grade, an octopus sailed off the seafloor and secured herself to a rocky outcropping off the coast of California. She was nearly a mile below the surface, thousands of feet past any tendrils of sun. But in the bright beams of a submarine, the octopus's edges glowed the reddish purple of a salted Japanese plum.

I know about the purple octopus because a remotely operated submersible watched her glide toward the cliff. The sub, which hailed from the Monterey Bay Aquarium Research Institute, had come to observe not just one octopus but the many *Graneledone boreopacifica* octopuses like her known to cling to this sea cliff. But she was the only one there, moving slowly toward the rock.

When the sub returned a little over a month later, it

found the same octopus—they could tell by the scars—latched onto the side of the outcropping, her arms coiled around her like suckered fiddlehead ferns, sealing in a newly laid clutch of eggs. When the octopus held herself close in this way, she was around the size of a personal pizza. Her large black eyes peered down into the abyss of the canyon beneath her.

The sub returned again and again to visit the mother octopus, who remained frozen in her vigil. She did not move. She did not eat. She shrank. Each visit found her paler, as if she had been dipped in milk. Her black eyes swirled in pallid clouds. Her pebbly skin hung loose from her body. The sub kept returning until it had seen the octopus eighteen times over the course of four and a half years, until one day it arrived to find the octopus gone. She had left behind a silhouette in tattered egg capsules still clinging to the rock like deflated balloons. This, the scientists understood, was a sign that her eggs had hatched successfully, freeing the mother octopus to die. Most mother octopuses lay a single set of eggs in a lifetime and die after their brood hatches.

The scientists who observed the octopus called her four-and-a-half-year brooding period the longest on record for any animal. In other words, no other creature on Earth had held its eggs close to its body and protected them for as long as she did; a story in Reuters called her "mother of the year" in the animal kingdom. The previous octopus record holder, *Bathypolypus arcticus,* was observed in captivity brooding for fourteen months, which seemed shattering at the time.

When I read about the octopus, I thought about sharing the article with my own mother, but I worried it might be too on the nose. I was hungry to learn everything I could about that mother octopus. I wanted to know how she chose that rock and how far she had to travel. And what her eggs felt like before being laid, if they were heavy, if they left an imprint on her body. And what else of the sea she'd seen up until then, and how she knew when it was time to leave the abyss, her familiar, dimensionless expanse. In the abyss, a body can move across three planes. In the abyss, where a human would sink, crushed by pressure and cold, an octopus can meander. It can roam and hunt and unfurl its eight limbs like a blooming flower.

Do female octopuses know what to expect when they're brooding? Does each mother learn about the vigil as she experiences it, wondering each day how much longer it will last? Hundreds of octopus mothers speckling a cliff, each starving, each alone. Or maybe the purple octopus, in her youth, passed by paling octopuses clinging to the edges of the canyon and recognized this would also be her fate one day.

More than anything I wanted to know why the octopus, with her big and alien brain, did not eat while she brooded her eggs. Surely she must have hungered. Did she have any inkling of the flurry of babies that might not make it if she strayed from her vigil to hunt or eat or stretch her limbs? I knew I was anthropomorphizing, and yet I couldn't imagine how a creature with a consciousness would starve for four and a half years without something like hope. What I mean to say is: I wanted to know if she ever regretted it.

The way I remember it, I first noticed my body sometime in middle school, after opening a Christmas gift—a trompe l'oeil of a shirt that pretended it was two layered shirts when, in fact, it was only one. As I tried it on before a mirror, I noticed my stomach, soft and round, pressed up against the cloth and peeking out from below. I felt ashamed for not seeing it earlier, for not paying attention.

The way my mother remembers it, I first noticed my body sometime in middle school, one day in the kitchen. She says I walked in and approached her, that I pulled up my shirt to expose my stomach and told her I was fat. She says this conversation is still etched into her memory, after all these years.

My mother, five feet three inches, rarely weighed more than 115 pounds. When she did, she said she was fat. When I was a child, she would tell me that when she was younger she weighed 98. She said this is when she was skinny. When my mother weighed 110 pounds, I weighed 115, and then 118, and then 124. I knew this because I tracked it every day, sneaking into her bathroom to step onto her digital scale. I would take off all my clothes and drop them in a heap beside the scale. I would close my eyes as the numbers scrambled. I think I kept them closed longer than I needed to, afraid to leave this unknowing. Sometimes, when the number disappointed, as it often did, I would weigh myself again, futilely repositioning my feet as if this could shuffle the mass of my body. But the scale did not change. So I would step off, shrink into the corner, put my clothes back on. Even then, I knew I would never be as skinny as my mother's worst version of herself.

When I was in high school, my mother and I developed a ritual. She would pull me into her closet, open bags of carefully folded clothing, and ask if I wanted it—the pants that no longer fit, the shirts that were no longer hip. And I would take the bundle to my room and try it all on, watching my hips spill out, my cinched body gasping for space. And I would return the bundle, say something like "It's not my style," and then a year would pass and we would do it all over again, my valiant squeezing, my mother and I deluding ourselves in different ways.

In the animal kingdom, there are two ways to be a mother. Some animals can reproduce multiple times in the span of a life, others just once. Humans, like most plants and vertebrates, have more than one chance to bear young. We can care for our babies and watch over them, and in doing so we increase their chances of making it to adulthood. We may even grow old with them. But creatures like octopuses have no such maternal privileges. Their single shot at reproduction produces hundreds or thousands of babies, stacking the odds that at least a few will make it out alive.

Octopuses brood all over the sea. In shallow-water dens, giant Pacific octopuses lay tens of thousands of tiny eggs, strung from rock like dangling hyacinths. The purple octopus lays fewer, bigger eggs, each the size of a large blueberry. If you lay only 160 eggs, only 160 chances that your young will survive, you must watch over them as long as you can. You must pour as much of yourself into making them as strong as you possibly can. After she lays her eggs, the mother octopus bathes them in new waves of water, doused with oxygen and free of any silt or debris. Her eggs need to breathe, so these baths are unending, kept up until the moment the eggs hatch. The purple mother octopus in Monterey Bay chose to lay her eggs on a sheltered alcove on the canyon wall just a few feet above the seafloor. The scientists noted how the crown of a rocky shelf above her shielded her eggs from unwanted silt. The spot was perfect, it seemed, and she must have known it.

Octopus eggs offer precious nutrients in the immense sea, meaning the octopus mother cannot leave her post to hunt. She survives on the stored energy of her body. She will never again see another place; this is her last view, enlivened only by the freer creatures that happen to pass through the icy waters. In the deep sea, these visitors are alien: fish with transparent faces and golden eyes, ghost sharks, tongue-red worms.

My mother immigrated to the United States in seventh grade. She moved from Taiwan to Hancock, Michigan, one of the snowiest cities in the snowy state. Hancock, Michigan, where it has been known to snow in June. Hancock, Michigan, where all her neighbors were tall, pale, and blond. My mother had come to Hancock to stay with my grandmother's sister, who wore her inky mop of hair straight down her back. Costume jewelry the size of beetles clung to her spindly fingers.

My mother spoke only Mandarin, so every day the kids at school reminded her in words she could not yet understand that she was different. Not like them. This was the first moment when my mother learned to want to be as American as possible. To have blond hair like her classmates, to have their blue eyes and overalls and long legs. She told me that it felt like she was an alien on a new planet. "You do what you have to, to survive," she said.

When my mother was pregnant with me, she gained 40 pounds—more than she had expected or wanted. When she went in for a checkup, all 40 extra pounds plus me, her doctor told her to stop eating all that Chinese food. "That doctor was a bitch," she said.

If my mom grew up wanting to be white, I grew up wanting to be thin. I sometimes wondered if I were full Chinese, not half, thinness would have come naturally. I never considered this obsession a disorder; eating disorders were for white women, said the movies and the magazines and the clinical papers. In front of mirrors, I squeezed the fat from behind my thighs to see how big my bones were, and if they were bigger than my mother's, I blamed my whiteness. I needed something to blame because every weekend when I saw my grandparents I saw my body disappoint them. When my grandma pinched the hammock of my arm and asked me if I'd gained weight, when my grandpa took one look at me and scoffed, "Big girl, too big!" I needed something tangible to explain to him why I was like this, so unwieldy. It couldn't have been my fault, because I had tried everything. Running every morning. Seltzer instead of snacks. Laxatives when I was desperate enough to feel my body mercifully, urgently hollowed out. But every time I tried to starve my body, I found I could not. I was too ravenous, too impulsive.

When the mother octopus first crawled toward the deepwater canyon, her body was purple. Her skin was corrugated in nubbly constellations. But her colors faded while brooding, her skin blanched white. She became the colors of her scars. As she brooded, her body became a beacon to anything passing by with a light, a gleam in the shadows.

There is technically no way of knowing if the purple-turned-white octopus ate anything in her fifty-three-month vigil, but there is no indication that she did. When the sub returned to visit, it observed spidery king crabs and vermillion-colored shrimp, common prey for deep-sea octopuses, skulking around the brooding mother. But the octopus never seemed to consider this emboldened prey as anything more than a threat to her cluster of young. When the crustaceans edged too close to the delicate clutch of eggs, she pushed them away with a flick of an arm.

During one of its many visits, the sub offered the mother octopus small pieces of crab with its robotic hand, manipulated by scientists on a boat thousands of feet away at the surface. But the octopus refused, not even willing to taste. The one examination of a brooding *Graneledone boreopacifica* revealed an immaculately empty gut.

When the running did not work, I asked my mother to put me on a diet. It was French, named after Pierre Dukan, a doctor who called obesity the twenty-first century's greatest "serial killer," who had his medical license revoked for commercializing his trademark diet and was sued for prescribing a patient an amphetamine-derived drug believed to have killed hundreds of people.

It was summer break, and I had nothing better to do than starve myself. The first phase of the diet allowed only lean protein, plus one and a half tablespoons of oat bran and six cups of water each day. In the morning, I spooned egg whites into my mouth. For lunch, wafer-thin slices of turkey pleated like an Elizabethan ruff. Before bed, a spoonful of oat bran, which I stirred into a glass of skim milk and drank, the dusty grains clinging to my throat. Soon I could eat non-starchy vegetables: kale, cabbage, and carrots, but no corn or potatoes. This was the diet I was to stick to until I reached 110 pounds, which I had decided was my goal weight. The number 110 was so roundly symmetrical, fluffed with an aspirational zero, and I insisted to myself I could do this; after all, I considered myself an overachiever in every other sense. Each day I woke up and felt weaker, a sign I took to mean Dukan was working its magic on me. Some afternoons, too entranced with my hunger to read, I nestled my body into the grass, eyes closed and limbs splayed like a sea star. When I closed my eyes, I didn't imagine food. I imagined my fat melting into the ground; like a whale fall, my bones would be the only things left.

I lasted a month on the Dukan diet. When I told my mom I was stopping, she asked if I was glad I had tried it, and I was, of course I was. When she left the room, I snuck two pieces of whole-wheat bread to my room and scarfed them down so fast I barely tasted them. This was the limit of my imagination, two pieces of whole-wheat bread. I lay back down in the grass but had lost access to my fuzzy dream state. I felt my blood rushing, my heart beating, my shame a new life force. I felt the blades of grass prickling into my skin. My body had become some alive and slippery thing I could not hold.

In the deep sea, everything starves. Space is depthless and barren here, life scarce, and meals few and far between. The water averages 40 degrees Fahrenheit, slowing metabolisms to a trickle and ensuring animals hold on to their fat as long as they can. The large creatures go weeks, even months, without eating in their aimless foraging. Giant isopods, lavender pill bugs the size of casserole dishes, can survive for two months between meals. The apple-sized white snail *Neptunea amianta* can last for three months. These stretches, not as grand as the purple octopus's, are a way of life.

All this starvation makes you smaller. The deeper you go, the smaller creatures become. Past two and a half miles, minuscule creatures—crustaceans called copepods and single-celled foraminifera—dominate the abyss. Bacteria teem. Two researchers realized their entire collection of deep-sea gastropods from the western North Atlantic—a trove of more than twenty thousand shells—was so small that all twenty thousand could fit inside a whelk shell the size of a fist.

So far removed from the light of the sun and the power of photosynthesis, deep-sea creatures depend upon the constant drizzle of marine snow—flakes of snot, poop, and disintegrated flesh from the world above them. Some flakes take weeks to reach the seafloor and grow as they fall, accumulating into white tufts. What is not eaten disintegrates into the ooze that carpets three-quarters of the deep ocean floor. At these depths, it is always marine snowing, always marine winter.

But the arithmetic of this deep-sea food web does not add up on its own. Even a constant snowfall of organic matter is not enough to sustain the vast communities living in the deep ocean. In 2013, scientists conducting a decades-long observation of marine snowfall on an abyssal plain off the coast of central California noticed three dramatic spikes in the data that solved the puzzle. The spikes marked windfalls of food—as fresh as it comes in the deep sea. The first was a bloom of microscopic, needlelike diatoms that erupted at the surface and plummeted to the seafloor. The second was a bloom of gelatinous salps, whose fallen bodies blanketed the seafloor in a silvery sheen. The third was a bloom of algae, which also sank in great waves and splotched the seafloor in brown fringe. Each of these feasts disappeared rapidly, binged on by creatures that had traveled to know, perhaps for the first time, what it meant to feel full. The scientists reasoned that this must be how deep-sea communities sustain themselves: long periods of restraint interrupted by serendipitous sprees of indulgence.

The offices of Trimm-Way Weight Center were on the second floor of a prim shopping mall that also boasted a State

Farm and a dry cleaner. My mother and I arrived early, and the receptionist gestured for us to sit on the couch, which was white, below a glossy black woodcut of a swooping crane and golden reeds. The print was clearly Asian inspired, and I imagined what my nutritionist might look like: black hair, poreless skin, a body like a blade of grass. But when the receptionist waved us into the office, I saw that my nutritionist was white, and her name was Karen. Her dyed hair was the color of a banana, and her thin legs sprouted from black stilettos with a platform like a brick.

My mother explained to Karen that I would like to lose weight. Karen looked me up and down and nodded. "You're lucky—you don't have to lose much," she told me, her teeth flashing. I felt both relieved and, unexpectedly, distressed. I hadn't realized a small part of me was hoping the nutritionist would tell me that, actually, I was fine. That, actually, I could just exist in my current body, and the real work was to love it.

But Karen was not a nutritionist. Karen was a weight-loss coach, and I stepped on a scale when she asked. She wrote my weight down in a small black book, and when I peeked I saw the names of her other clients, all of us plotted in a tragic grid. I stepped on another, more intricate scale, which Karen explained would measure my body fat by shooting electrical currents up my leg and across my pelvis. Karen typed all these facts about me into her computer and printed out a packet, which she handed to me. The front page had a typo in my name, Sabirna, and as I read that name I imagined my skinnier doppelganger.

Karen told me I could eat three meals of 300 calories a day as well as a 100-calorie snack. For breakfast, I ate three turkey sausages (100 calories), a glass of milk (90 calories), and an apple (100 calories). For lunch, I ate cottage cheese (100 calories) and grapes (100 calories). For dinner, I ate chicken breast (200 calories) and a vegetable, maybe broccoli (100 calories). All day I craved snacks horrendously, so much that I sometimes skipped dinner to eat 100-calorie packs my mother bought me, crinkled envelopes of wafer-like Chips Ahoy! and cardboardy Oreos and shrunken Wheat Thins. I chewed these until they became mush in my mouth, knowing that after I swallowed there would be no more.

When news outlets wrote about the purple octopus, they fixated on the numbers associated with her life. Fifty-three months, four and a half years, 4,600 feet below the surface. This is how the dead octopus became statistically significant, a viral darling. When journalists wrote about her, they marveled at her body's great and terrible capacity to stay alive while starving itself to death. *Graneledone boreopacifica* is one of the most abundant octopuses in the eastern North Pacific, meaning there are untold other octopuses sitting on their eggs for four and a half years or longer, whose sacrifices we did not happen to see.

In the Monterey canyon, the black-eyed squid *Gonatus onyx* carries her thousands of eggs in her arms as she swims. The eggs cling together in an enormous cluster and twinkle like a disco ball. Every thirty seconds, the wine-colored squid extends her arms to flush water through the egg mass, refreshing her babies with oxygen. Black-eyed squids are agile on their own, able to jet quickly away from whales, elephant seals, and other deep-diving predators. But a mother squid's shimmering mass of eggs weighs her down, makes her slow and bulky. She still carries her babies, for six to nine months before they hatch. When they do, the mother squid dies; like the purple octopus, she has not fed for months.

Elsewhere in the deep, the giant red mysid *Gnathophausia ingens,* which resembles a shrimp, carries her eggs for approximately one and a half years. She, like the others, does not feed. She shrinks to a fraction of her size, losing body mass steadily as she drifts in the blackness. Her eggs require 61 percent of the energy she has accumulated over her lifetime, meaning she gives more of herself to her babies than she does to herself. When her eggs hatch and the larvae swim away, she dies.

Scientists only know about the brooding habits of the purple octopus, the black-eyed squid, and the giant red mysid because they have observed them. Subs often encounter creatures in the deep by chance, capturing fleeting moments in strange and secret lives. There may be many more creatures that mother this way. Thousands of other octopuses and squids and mysids, starving themselves in the abyss.

During my summer of Trimm-Way, there were days when my hunger became so great that I inhaled food—five bowls of cereal, three bags of popcorn, an entire box of Wheat Thins. When I was done I would lie down in our yard, eyes closed, stomach in pain, dreading the lie I would have to tell at my weekly weigh-in. Sometimes I chewed over the trash can without swallowing, spitting out orange mouthfuls before they touched my throat.

Some weeks I stopped drinking water before walking the half hour to Trimm-Way. I walked slowly, dizzy, each step a drip. One day the sidewalk blurred and I fell. I walked myself to a bench, where I closed my eyes and sat. I felt weak, deliciously frail. I was fifteen minutes late to my weigh-in, but I had lost 4 pounds. Karen beamed, hands clasped as if in prayer, oblivious that the weight I had shed was water. As soon as I left Trimm-Way I bought a family-sized bag of Cheetos at the drugstore and ate the entire thing on the way home.

During my first year of high school, I overheard a girl named Alex talking to her friend. "I would give anything to be anorexic," she sighed, taking her burrito out of the cafeteria microwave. "But I just don't have the discipline." I felt a pang of agreement, or maybe it was hunger.

Narratively speaking, motherhood may seem to be the climax of the female octopus's life, the grand finale. It is the last thing she does before she dies. Male octopuses die soon after mating—sex a climax in every sense of the word—but female octopuses live long enough to brood the eggs. It is an extension of life, but also of labor. This arrangement, evolutionarily speaking, seems to me a raw deal.

People who care for captive mother octopuses have witnessed the animal's final moments, often called a death spiral. Some hurl themselves against the walls of the tank. Some rip off their own skin. Some even begin devouring themselves, tearing into the tips of their tentacles like they would a crab. That last image has seared itself into my mind. I wonder how those octopuses like the taste of themselves, their first meal after so many months of starvation. Do they savor it?

Scientists investigating this maternal death drive discovered the octopuses were simply obeying their optic gland, which runs between their eyes. In 1977, a psychologist removed the optic gland from fourteen female Caribbean two-spot octopuses, between the two iris-blue spots on either side of its head. When the glands were removed and the octopuses woke up after surgery, most of them abandoned their eggs. All of them began eating again, doubling their body weight from that shrunken, brooding state. Most of the mother octopuses doubled their life spans, living months after the moment the scientists had expected them to die.

This discovery was accidental. The scientist had removed the gland in a female octopus only because he was practicing for the real thing, the same optic-gland surgery on a male octopus. The scientist wanted to know how male octopuses would behave after these glands, which help control sex and reproduction, were removed. He knew all female octopuses died after brooding their eggs, so, he reasoned, if things went wrong during surgery, it wasn't like he killed something that would have lived. So here is another way to think about the purple octopus. If mother octopuses are condemned to die soon after their eggs hatch, the longer she broods, the longer she lives. It is true that the purple octopus shattered records for her brooding period. If the purple octopus broods, like most other octopuses, for a quarter of its lifetime, then she may be the longest-lived cephalopod that we know of. The oldest octopus in the world, the oldest of those supernaturally smart creatures who live in brief, dazzling spurts, some species less than a year. It seems a shame that an animal able to sense so much of the world occupies it so briefly, spends all of it at the bottom of the ocean, in darkness, at temperatures near freezing. But still, she lived.

Sometime in college, when I began to wobble toward "better," I saw that my mother was not. She still occasionally called herself a fat pig. But I had always been afraid to talk about it with her. How can you stage an intervention for your starving mother, especially when you worry you might still want to starve yourself?

The first and only time I asked my mom about her eating, she was on the couch watching PBS *Masterpiece Theatre*. I started by talking about myself, how my body had repulsed me for so long, how I was not sure I was entirely better, how I was hopeful I could be. A long silence later, she asked me: "Are you saying it's my fault that you're like this?"

"No, not that. I'm just—I just want to tell you that I think you, that you might be a little too thin, that you're not a fat pig," I hedged, uncertain, apologetic.

My mother says she does not remember this conversation.

I realize now that my mother's wish for me to be thin was, in its way, an act of love. She wanted me to be skinny so things would be easier. White, so things would be easier. Straight, so things would be easy, easy, easy. So that, unlike her, no one would ever question my right to be here, in America. I just wish I could tell her I've been okay without those things, that I've actually been better without them. I wish she would stop wanting those things too.

There is no turning point, no clear moment when I started feeling good in my body. I know that when I started dating people who are not cis men, I learned to revel in queer bodies and the endless and inventive ways we crease into ourselves. When I desired these bodies and the people who inhabit them, I began to see how my own body could be desired, not just by others but also by me. Years later, in a wry twist of queerness, when I begin to wish my chest and hips were smaller, my old hatred burbles back to the surface at a different slant. This time the wish feels tacky, because I know an androgynous body can exist in different sizes, because I know narrow hips are not a universal end goal, and yet my childish envy still seeps out. I predict I will always be in negotiation with my body, what it wants, and what I want of it.

In 1998, a different female *Graneledone boreopacifica* was collected by a submarine volcano off the coast of Oregon. The sub found her on the side of the caldera, reached out with a mechanical arm, seized her by her mantle, and placed her in a five-gallon bucket. The octopus fought back — "reacting vigorously," the researchers noted — perhaps becoming a pinwheel of tentacles, beak widening, suckers grasping the hard plastic of the bucket.

In the lab, the scientists slipped their hands inside the octopus to find her torn apart. Her digestive gland, part of her intestine, and her ovary had ruptured during the process of her capture, which released the contents of her intestine into her body. The hard parts of the animals she had eaten spilled out like confetti: bristles and jaws of rag worms, crushed whorls of deep-sea snails, vent limpets smashed apart like puzzle pieces. Researchers were stunned by the contents of her stomach; they had no idea that octopuses swimming at these depths were capable of crushing a calcium carbonate shell and swallowing it. They'd assumed that the soft-bodied octopus preyed on soft-bodied creatures. They never knew she could prey on something hard or sharp. Altogether, she had consumed at least seventy-six creatures. Before she died, the octopus had feasted.

At some point, after the running but before the diets, my mother took me and my sibling to visit her college. We were only an hour's drive away, she said, and she couldn't believe she hadn't taken us before. The campus passed in a blur as we sped by green lawns, picturesque libraries, brutalist buildings. We stopped at a hot dog shop with a blue-and-gold sign featuring a dapper wiener leaning on a cane. When my mother told us she ate here every week, I thought she was joking. The menu had only hot dogs, fatty sausages, and sodas, things I've never seen her touch. She told us to order whatever we wanted. I said I wanted to eat what she used to eat, so she ordered us enormous brats, slathered in sauerkraut, relish, and mustard. The dogs shattered in our mouths, fat and sauce leaking down our chins. I asked if my mom wanted a bite, and she shook her head. She watched us finish, wiped our dripping faces with napkins, and walked us to get frozen yogurt right down the block, another old meal, and we ordered spires of plain yogurt teeming with sprinkles. My sibling and I savored them in the car, so slowly we ended up sipping on marbled puddles of sugar. I licked my spoon and watched the streamers of light blinking on the bridge back home. I closed my eyes and imagined myself as my mother, my stomach my mother's stomach, back when she was young and tasted whatever she desired, back when she feasted.

My Grandmother
and the Sturgeon

The Chinese sturgeon resembles something from a past world, when scaled giants roamed the earth and the continents still clung together. By most accounts, it is ugly: leathery body outlined by rows of mud-brown armored plates and chin fringed by four fleshy barbels. But up close, the sturgeon's skin is almost beautiful, an opalescent sheen of yellows, greens, and grays dappling a muscular body.

The first sturgeon appeared around two hundred million years ago, an era when the seas around Pangea teemed with ammonites and the ground shook from the footsteps of 80-ton dinosaurs. When the asteroid wiped out the dinosaurs, the great fish survived. Some scientists call the sturgeon a living fossil. There are historical accounts of sturgeon as long as 16 feet and weighing half a ton. The fish do not grow that big anymore, not because they have changed but because the world has.

In this century, I am sitting on the floor with a handful of tourists, watching a sturgeon paddle languidly behind a wall of glass in Coney Island, New York. The strip of boardwalk is the last place in the world one might expect to see the big fish, an aquarium squeezed in between a roller coaster, handball courts, and a luxury shower door dealer. But they are here, five of the prehistoric-looking beasts, weaving apathetic paths through water they share with sharks, rays, and glinting metallic fish. They look like ancestors, conjured into a modernity in which they seem out of place.

At a distance, sturgeon could almost pass for sharks, with their leathery skin and sinister pinprick eyes. But the longer you stare, the more primordial they appear. Their range of mountainous scutes and chin bristles jut out like stalactites. They make sharks look laughably modern, all sleek lines and aerodynamic contouring and unfeeling marble faces. The sharks move with dogged, torpedo purpose, tails swishing back and forth as they circle around the perimeter of the tank. But the sturgeon glide aimlessly, with an ossified kind of grace, as if they know they are lost, so far from home and what they once were.

My grandmother grew up believing she was ugly because everyone told her so. A friend of her father's, their wealthy neighbor's sixth concubine, always told my grandmother she was ugly, even for a five-year-old. My grandmother felt proud to be loved by her father despite being so ugly.

Her father worked as a banker in the city. Her mother managed their home and my grandmother's siblings. (One sister, the fourth girl, had been sent to live in an orphanage in the countryside because their mother believed the child was cursed.) Her parents and siblings all slept in one room at the center of the house. Her own grandmother, whom she called Po Po, slept in a separate room, bedridden from years of eating opium. After school, my grandmother kneaded her grandmother's legs to relieve the pain and keep her blood flowing. My grandmother never forgot the strange smell in that room, which lurked under blankets and in drawers.

Within years, Shanghai fell to Japanese occupation. Over the following years, the Japanese armies bombed, massacred, and swallowed more and more of China—Nanjing, Guangzhou, railways in the Yangtze valley. The government of the Republic of China moved their capital inland, to the city of Chongqing.

When my grandmother was nine, Po Po died, freeing her family to flee Japanese-occupied Shanghai. They resolved to leave but told no one of their plans, unsure of who among their neighbors could be a spy. They believed they could make it to Chongqing in less than a month. It took six.

Though most sturgeon spend their adult lives in the sea, they are born in fresh water. Like salmon, ocean-dwelling sturgeon must swim upstream to reach their breeding grounds. For millions of years, adult sturgeon in China made a 1,900-mile journey about every four years, swimming up rivers to reach one of many spawning sites inland. Females made the trek as early as thirteen years old and males as early as eight. In summer, the fish breached the continent through the mouth of the Yangtze and swam for months while fasting, gradually depleting the reserves of energy they had built up for the occasion. The sturgeon spawned in a stretch of more than 375 miles of river that promised a more sheltered childhood than the void of the open ocean. In late fall, when red and orange leaves spiraled into the Yangtze, the sturgeon returned to the sea.

When times were good, the sturgeon could lay hundreds of thousands of eggs apiece, blanketing the bed of the river. Out of every million eggs, only six baby sturgeon would survive, escaping the mouths of hungry predators, to grow into larger fish. For sturgeon fry, survival in the Yangtze has always been a game of chance; it is far easier to die than it is to live.

Sturgeon still attempt to make this great migration each year, but they're now interrupted by a cascading series of dams that blockade the river. The Gezhouba Dam came in 1981 and sealed off the upper reaches of the Yangtze, severing the sturgeon from all but one of their spawning grounds. Other river dwellers, including the baiji river dolphin and Chinese paddlefish, have gone extinct, unable to survive the dams. But the Chinese sturgeon remains, a dogged survivor in China's largest river.

The Yangtze River originates in the Tibetan Plateau, which stands three miles above sea level. Some call it the Roof of the World. Here, streams of glacial meltwater converge into a river, which carves through the Tibetan highlands and tumbles down the mountains; slicing through gorges, fed by tributaries, it surges across Sichuan province, widens in the industrial ports of Chongqing, and continues on to Hubei, where it flows between the sheer rocks of the Three Gorges, great limestone pillars, before continuing onward. Hundreds of years ago, wild monkeys clung to the green trees of the cliffs and made their presence known by their loud, ceaseless cries.

In Chinese mythology, there is a legend about carp that climbed a waterfall on a mountain of staggering height. An ancient hero cleaved the summit of the mountain in two, opening up a gateway so the river could career off the cliffs. Each year many fish attempted to swim against the mighty current, and many failed. But a few fought against the surge of the river to take a final, breathtaking leap over the open gate at the head of the waterfall. Before their fins broke the surface of the water on the other side, the fish found their soft, plump bodies transformed into something slithering, and the slick mucus of their scales hardened into jeweled pebbles of skin. That is to say, they became dragons.

For hundreds of years, like the carp, Chinese sturgeon

swam against the current of the Yangtze. After the Gezhouba Dam was built, the great fish could be seen flinging themselves at the dam, attempting to cross. They hurled their bodies into the concrete and steel, again and again. How had a passage that was once open suddenly become a wall? Many were injured, retreating to the sea with bruised and battered snouts, their unused ovaries shriveling up. Others died, their bodies sinking at the base of the dam, almost like an offering.

To travel on the river through Hubei province when fleeing Shanghai, my family hired two small houseboats, each carrying one captain and one crew member. The boats had no engines and were powered only by rowing, and so my family traveled the river at an achingly languid pace, a nightmare of a leisure cruise.

My grandmother slept in the belly of one boat alongside her family, stacked like sardines on shared padding. They ate one meal a day, always rice or congee, cooked in soup that was more water than broth. Occasionally the captains would return from the village with a feast of some vegetables or an egg, maybe two. Eggs were always given to the grown boys or to the youngest children, never to my grandmother.

The boys rowed and everyone starved. My grandmother sat on deck and scoured the banks for signs of life, spotting empty houses and ghost villages, families fleeing the fighting. Sometimes they passed Japanese troops stationed by the side of the river who would call their boat over and take the boys to do manual labor. Each time this happened, the women and girls waited on the boat, remembering that soldiers often do not return the boys they steal. Each time, the boys returned.

Gradually, like twilight, the river blushed pink. Soon enough the bodies came, rarely intact but always recognizably human. First, an arm or leg. Later, a torso. At one point, unforgettably, a head. They all wore the clothing of farmers, country people. The dead became routine, grisly apples bobbing downstream. They arrived in batches, indicating that the Japanese had just taken another village.

When the river brimmed with bodies, the captains knew to pull the boats over and hide. Sometimes my grandmother could hear, beyond the emptied villages that lined the river, the howled commands of Japanese soldiers. Whenever the boats strayed close enough to hear the voice of a lone man, the crew rowed faster. Ammunition could cross the river like a bird. One day, the bombing stopped.

With all the farmers dead, the eggs ran out. After the last grain of rice had been swallowed, the crew pulled the boats over and docked them at a bank. They had grown too hungry to row, with no village in sight. They had no rice, no flour, nothing that could make a meal. The kids had no energy to cry anymore. Everyone lay flopped on the decks, waiting to die.

The Chinese sturgeon is dying out, and it is not alone. All but four of the world's twenty-seven species in the family Acipenseridae hover close to extinction. So the sturgeon are dying, in lakes and rivers and oceans all over the world. These giant fish survived the asteroid and the Ice Age and so much more only to be wiped out by cosmically puny obstacles: our dams, our boats, our chemicals, our taste for caviar.

In his essay "Sturgeon Moon," the fisher John Cronin asks us to consider the sheer magnitude of the animal's heritage in new units. If we translate two hundred million years into a twenty-four-hour clock, we have taken less than one-tenth of a second in the last minute of the last hour to imperil every single subspecies of sturgeon on the planet. Such is the reach of their history and our power to destroy it. Today, there may be fewer than one hundred Chinese sturgeon capable of returning to the Yangtze each year to breed. Sometime in the next ten to twenty years, scientists predict, the Chinese sturgeon will go extinct in the wild.

After the rice vanished, my family remained on the river, docked, prepared to waste away. Suddenly, along the bank of the river, a Japanese soldier approached their boats. The captains hesitated but allowed the soldier to board. My family did not speak Japanese, but hunger is easy to communicate with your hands. The Japanese soldier gestured back with his bayonet and left. My family did not know whether to weep or rejoice. Either way, they were dying.

A few hours later, the soldier returned, carrying something heavy. He boarded the boat and dropped on the deck a burlap sack as big as a torso, gestured toward the captain to pick it up, and stepped back onto the bank and disappeared. The sack contained rice, thousands and thousands of milky white grains spilling into one another like the froth of a breaking wave. Everyone was shocked, still reeling from a newfound realization they would not die just yet. The captains thanked the soldier again and again. That night they ate congee, diluted with river water.

While my grandmother lay starving on that boat, young Chinese sturgeon lurked at the bottom of the Yangtze, foraging for food in the scarlet murk of war. They may have tasted the river's newly sour, ferric tang.

In 2012, the city of Chongqing saw the Yangtze blush once more. News outlets described the color as tomato red, scarlet red. Scientists speculated the hue came from sediments, upended by heavy rains upstream. Or perhaps, they said, the red tide came from a sudden algal bloom, the natural reaction of the collision of fertilizer runoff and nutrient-rich waters. But the redness did not raise real alarm, as rivers flush red in China more often than one would think. Twice in the last decade, accidental pollution from dye factories has turned rivers in China various shades of red. Reds the color of beets, fire engines, wine, cinnabar, rust— anything but blood.

To swim in the Yangtze now is to bathe in synthetic compounds. Industrial and agricultural pollutants unspool into the river, runoff from city drains and industrial sites. The most dangerous are compounds of triphenyltin, or TPT, a biocide that fishers use to coat their ship hulls and farmers spray in paddy fields to kill off golden apple snails. These chemicals have the power to reshape a fish.

The Chinese sturgeon can accumulate more TPT in their livers than can other fish. These poisoned fish lay eggs that hatch into deformed fry. Their larvae have just one eye, or no eyes at all. They cannot see how the river they inhabit is not the same as the one their ancestors knew well. Their spines bend like used paper clips, arching in hard angles that make it impossible to swim.

The Chinese Sturgeon Museum, located on a small island in a tributary of the Yangtze, is the only museum devoted to immortalizing the giant fish, as if in anticipation of its extinction. It is an extension of the Chinese Sturgeon Research Institute, a facility tasked with restoring the sturgeon's population. Inside the museum, you can see a variety of sturgeon preserved in varnish or coiled inside preservative fluid. Living sturgeon swim tight laps in a shallow pool, tinged green with algae.

Sturgeon have more real estate at various breeding facilities by the river, all of which hope to restore the sturgeon's population in the wild. At the Yangtze River Fisheries Research Institute in Jingzhou, artificially bred sturgeon grow up in rows and rows of tanks. China Three Gorges Corporation, the company behind the Yangtze's biggest dam, has released batches of baby sturgeon into the river since 1984. But hardly any of the millions of sturgeon fry dumped into the river over decades have survived, and dumping even more of them does nothing to ensure they will make it into the ocean. It is easy to give these babies away but impossible to be certain they will survive.

Before my grandmother's family arrived in Chongqing, five months late with only the clothes on their backs, they found themselves lost, separated from the rest of their group in the darkness. As they huddled in a deserted temple, her mother spoke to her children in whispers. "If we are discovered by enemy soldiers," her mother said, "I will kill you first, I will kill you before the soldiers can." What my grandmother's mother meant was: *We have no way of escaping the soldiers and their metal bayonets. If we have to die, let it be gentle.* My grandmother and her brother wept, silently, in the darkness. Years later, after my grandmother had grown, she still woke up in the middle of the night, remembering the time her mother promised to kill her. It took her years to understand it was a promise filled with love.

Since the Gezhouba Dam spliced the Yangtze, the sturgeon's spawning grounds have grown eerily empty. Not only have the sturgeon been cut off from their breeding grounds but also the construction of new and bigger dams farther inland has raised the temperature of the water, conjuring warmth that now lingers longer, from late summer into fall. The fish, which find it hard to breed in warm waters, delay their spawning until the river cools, giving them much less time to breed. And all the construction required to build these dams—as well as river traffic—conjures a deafening underwater roar. If my grandma wanted to retrace her childhood migration, she would be disoriented beyond belief, not by a dam but by a land rendered unrecognizable to her, huts and lanterns replaced by high-rises and the beams of electronic billboards. The place my grandmother now knows best is a city in California named for a Catholic saint. There are no rivers here, only creeks.

Over this past year, my grandma's memory has begun to blot out. My mother was the first to inform me, warning me over the phone that I should not act alarmed if my grandma lost her train of thought, or if she began speaking to me in Mandarin. When I speak with her on the phone, our conversations rarely last over five minutes. My questions seem to strain her, and she has no new questions for me except for when I am coming home, if I am safe, if I have enough to eat. When I do come home, I find her frailer, stubborn, still insistent on driving even though we ask her not to because she no longer remembers to signal. She speaks in Mandarin more and more, and I cannot understand her. Sometimes it feels like she is returning to China, and I wonder if she is also returning to her memories. I am afraid to upset or strain the tether she holds to the present, to America, to the only part of her life where I exist.

When I am sitting on the floor before the sturgeon tank at the New York Aquarium, I cannot help but think of her. These sturgeon are from the Atlantic, specifically the Hudson River, brown facsimiles of the fish I really want to see. I imagine them growing old in this tank, their roe preserved and flesh unopened in exchange for years of their endless circling. I wonder if they remember the waters they came from. Later, on the train home, I read in a press release that the sturgeon were bred in a facility in Maryland and have only ever known life in a tank. Maybe this life, insulated from the fishers and pollution and other dangers of their wild home, isn't so bad, I think, my body nestled against the steel bars that stand between my seat and the subway doors, which gape every few minutes like some giant creature breathing. I close my eyes. I remember the twin shadows of the sturgeon wrinkling over the glinting sand of the tanks. I feel very close to and very far away from home.

How to Draw a Sperm Whale

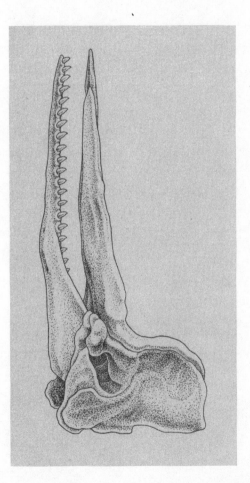

In the spring of 1998, off the coast of Nova Scotia, something nearly impossible happened to a blue whale. Young and healthy and small for a blue at 66 feet, he was feasting on eruptions of krill when he heard a bellowing, unnatural roar. Perhaps he swerved to avoid it, but he was too slow, and the ship's propeller sliced easily into his jaw, fracturing the bone and leaving behind a chevron-shaped cut. Soon he heard another roar, in the distance until it wasn't, and suddenly a second ship, a red tanker called *Botany Triumph*, hit him. The oblivious tanker continued on its course and carried him on the lip jutting out of its bow, an accidental Viking funeral, his body draped under just two feet of water and his blueish tail waving at the surface. Held there, somewhere in transit toward New England, the whale died.

At the time, there were fewer than 10,000 blue whales in the oceans, and the unnatural loss of even one blue whale in the western North Atlantic would threaten the population's recovery. Scientists estimate that we killed 360,000 blue whales in the first six decades of the twentieth century. I cannot begin to understand this loss—what a world with 360,000 blue whales looked like—just as I cannot understand the vastness of the blue whale, whose tongue weighs as much as an elephant, or how any human might believe themselves powerful enough to kill one.

The dead whale was discovered in the Rhode Island Sound by the captain of a smaller pilot boat, who radioed someone on the *Botany Triumph* to say the ship had hooked something on its bow. The ship came to a halt and reversed, nudging his body off the lip. The crew watched intently, perhaps praying, against all odds, for some sign of life. Some had tears in their eyes. But the whale slid off like a limp slug and began to float.

Four days later, the whale was towed to shore. A day after that, he was dissected.

Marine Mammal Stranding Report
Discovered on the bow of the *Botany Triumph*
Genus: *Balaenoptera*
Species: *musculus*
Condition: Moderate decomp

The National Marine Fisheries Service, which governs the bodies of whales in the US that strand in inconvenient places, donated the whale's skin to a lab in Belgium, a hunk of his blubber to Texas A&M at Galveston, his left eye to the Mystic Aquarium in Connecticut, his ear bones to the Woods Hole Oceanographic Institution, and his eight-foot larynx to Mount Sinai School of Medicine in New York, where it would be kept in a walk-in refrigerator. But the biggest part of the whale—his skeleton—went to the New Bedford Whaling Museum in Massachusetts.

He now hangs near the entrance of the museum, polished bones looking more like a chandelier than a body. His skull resembles an enormous beak. His flippers look almost like hands, long digits that would have been hidden under his skin. His spine terminates in vertebrae so small they look like apples, his lobed tail, called a fluke, nowhere to be seen. They call him KOBO, or King of the Blue Ocean: a name courtesy of the local sixth-grader who won the dead whale naming contest.

I first came to the New Bedford Whaling Museum to draw this whale and the many others—real and depicted—interred in its walls. I was in college and had enrolled in a painting class called Illuminating the Ocean Deep, which I thought was about whales. I soon learned the class was less concerned with whales than it was with whaling, the systematic hunting and harvesting of the animals that brought human populations to the verge of unimaginable prosperity and whale populations to the brink of extinction.

I had applied to take this class for two reasons. The first:

I was writing my senior thesis about whales, a sprawling and unfocused project that was more concerned with describing whales—living, eating, breathing, beaching, dying, and dead—than saying anything about them. The second: my first girlfriend, M, was taking it. Before the class, M knew how to draw whales and I did not. After the class, I was in love with M and they were not in love with me.

"Necropsy," the word, was invented to distinguish between the act of examining dead humans from that of examining dead animals. An autopsy would determine the cause of death of any animal, human or not, until the early 1800s, when a French doctor proposed "necropsy" for the nonhuman. All of us still die, but now we humans are autopsied and whales necropsied. Our exams involve the same process (opened up on the table) and end goal (to determine cause of death), but we are spared, at least etymologically, the suggestion of our corpsehood.

The word "necropsy" breaks down to "death" + "seeing"; "autopsy" to "self" + "seeing." When I first learned about this etymological split, it seemed silly, even redundant. It reminded me of all the ways we shoehorn distinctions between ourselves and other animals, often harming both of us. But I understand now that an autopsy can be an uncanny act of prediction for the dissector: one of many ways to go, a possible future. When biologists necropsy a whale, I assume they are not reminded of our individual threat of cancer or the universal threat of a car crash. They do not subconsciously compare their arms to a whale's fin, their teeth to bristly plating of baleen.

The Woods Hole Oceanographic Institution has compiled an exhaustive introductory guide to marine mammal necropsy. The guide suggests you obtain as much background as possible before examining an animal: the time and date of discovery, the environmental conditions, a record of its trauma. The guide also offers a bulleted list of tips to keep in mind during the examination: Develop a routine, and be objective. Document everything. Understand and acknowledge confounding variables.

The guide is surprisingly forgiving to first-time coroners. Describe what you see, smell, feel, and hear, it advises. Do not worry too much about using all technical terms. "Learning the language will come with time. Write it how you see it." For those, like me, unfamiliar with the terms, the guide also contains a glossary. Caseous: resembling cheese. Cyamids: whale lice. Ectasis: a widening. Fusiform: spindle shaped. Stellate: arranged like a star. Tortuous: having many turns, winding or twisting. Peracute: very acute, violent.

In the United States, every stranded marine mammal must be examined for traces of human interaction. Abstract charts filled with geometric markings offer clues about the killer. A row of ellipses indicates twisted twine. A series of diamonds, entanglement in a net. Long slashes or short crescents suggest a propeller was involved. The guide includes a list of questions and answers to help you identify abnormal lesions "found grossly," which means what you see with your own eyes, without the help of a microscope.

Where is it?
Dorsal, ventral, lateral, medial, proximal, distal, cranial, caudal, anterior, posterior.

What is the shape?
Round, spherical, ovoid, crescent, nodular, conical, lobular, tortuous, discoid, bulbous, sessile, stellate, reticular, fusiform, irregular, loculated, branched, amorphous.

What does it feel like?
Wet, dry, tacky, hard, firm, soft, friable, gas-filled, viscous, gelatinous, gritty, resilient, rubbery, granular, flaccid, depressed, raised, smooth, rough, nodular, grooved, crusted, spongy, thick, thin.

Though they have a precise application here, the questions sound eerily universal, something you could ask to describe anything: a fruit, a cloud, a bruise; what it is like to fall in love, to be wrong, or to die. The answers could be combined in a seemingly infinite number of ways yet always will result in something dead.

Necropsy Report: A Relationship

Time of death: Sometime in March—it may have been the morning
Location: New England
Sex: Not for a long time
Length: Half a year, depending on when you start counting

Submitted: A Trader Joe's bag of artifacts, including handmade paper, a case of art supplies, a bulbous dildo in a ziplock bag

History: You and M first saw each other in the library and officially met in a queer theory class (a cliché, obviously). You were ghosted that summer by the first girl you slept with (tacky!), a separation that left you depressed, unsure if you were in love, or if you were gay or straight, because you, young idiot, were under the impression that sexuality could exist beyond the binary for everyone but you. You felt afraid to return to dating men, that it would prove your first girlfriend right. You were also afraid to date people who were not men, that perhaps you were not queer but only wanted to be.

When you kissed M for the first time, you felt friable, like the contents of your cells had broken free of their membranes and were misting out of you. It was, for you, an infatuation that made time apart unbearable and time together infinite. It was, for M, likely something else.

External examination: There were no signs of any deception. You almost wanted something like that as an excuse, because such ends felt less painful than the idea that one person had simply lost interest. There were reports of pre-existing conditions: incompatible interests, incompatible desires, an imminent graduation. The proximate cause of expiration is still unknown.

The immediate cause may have been one last fight, in which both individuals accused (one more than the other) and both individuals cried (one more than the other). At some point, you turned to begging, asking M if there was any way to make them stay. The fight, ironically enough, occurred on Hope Street.

Despite its deadly appearance, a harpoon was not meant to kill a whale but to capture it on a leash. Its iron head, fitted with a variety of hooks and barbs as intricate as sailors' knots or as simple as a clover, clung into blubber and held fast. When the whale was just a few feet away, the harpooner plunged the shaft of the weapon into the whale's back and the crew rowed hard. The now-snagged whale might thrash in pain, might even dive, pulling the line of the harpoon so fast that it smoked. The men let the whale tire itself, then drew their boat closer to plunge the weapon, repeatedly, into the whale's softest spots: heart, eye, lung. When blood gushed from the blowhole, the men knew the fight was nearly over.

In the late 1700s, hundreds of ships sailed out of American ports in pursuit of whales at a scale unprecedented in these waters, their crews scanning the seas for distant plumes, a sign that, somewhere near the horizon, a giant was taking a breath. A dead whale could be farmed for a number of products—bone carved into corsets, teeth fashioned into the crowns of walking sticks, baleen bent into hoop skirts and umbrella ribs. In most cases, the most valuable part of a whale was its oil, which brought light to streetlamps in New Bedford, the United States, and Europe. Whale oil lit everything from candles to lighthouses, and sperm whales were unlucky enough to have the most desirable kind of substance, a liquid wax called spermaceti, sloshing around in their heads. Spermaceti burned smokeless and scentless, creating candlelight that seemed plucked from a sunbeam rather than scooped from the bloody hull of a whale's head. In 1857, New Bedford gave itself a motto in honor of its trade: *"Lucem diffundo,"* or "I diffuse light."

But its reign as a whaling capital was brief. By the 1860s, the industry began to falter. Pennsylvania had struck petroleum, which diminished the need for whale oil. In Europe, the advent of mechanized whaling technology allowed Norwegian whalers to capture and kill whales at an unprecedented pace, including finbacks and blues—species that had long evaded American whalers. Today, New Bedford is working class, and its fishers deal in sea scallops, mollusks that cannot kill a man but do not summon the glory of a whale.

We came to the museum for six weeks, our purpose in the class description to "contemplate the tenuous line between the pursuit of profit and the destruction of that which we hold most sacred." When I signed up for the class, I thought this contemplation would look like sketching whales. But the museum had no whales, only parts of them, disassembled and strewn in glass boxes. In one, corset busks made of whalebone. In another, a prominently displayed sperm whale penis, upright and desiccated and looking like a parsnip. Mostly there was scrimshaw, intricate carvings of whales or ships on teeth wrenched from a dead sperm whale's mouth. The crammed alleyways of the museum's archives held even more. A fetal sperm whale dunked upside down in ethanol, its tiny mouth eerily curved into a smile. A mauve-colored pair of cufflinks made out of the pupils of a right whale. I held them up and stared at them, musty orbs on tarnished silver backings, and sketched them. My inexpert hand rendered the pupils indecipherable—easily mistaken for peas or rocks—so I wrote "right whale cufflinks" and drew an arrow pointing to the balls.

Necropsy Report: A Relationship (*cont.*)

History: Things started going wrong, you thought, right around the time the two of you started living together. It was supposed to be temporary, to save money. And at first it was easy, close to perfect. You slept together in a bed pressed up against the window, the winter sun sliding between the slats of the blinds, and woke up together to the erratic clanging of the radiator. You marveled at the sudden intimacy of witnessing someone's daily routine: how they made their eggs, brushed their teeth, breathed at night. All of it was new to you, and to your college self, new things seemed objectively hotter than old things.

Internal examination: Maybe it was the unwelcome deluge of logistics required to live as an adult for what seemed like the first time: pushing a sessile car up a snowbanked driveway; mopping sewage spilled, miraculously, out of the bathtub; tinkering with a radiator using the remaining half of an IKEA tool kit.

Maybe it was the unanticipated anxiety of private moments. At night when the two of you got ready for bed, you would wash the concealer off your face, turning the sink water a gritty brown and inflaming the nodular bumps of cystic acne on your cheeks in the bathroom light. Once in the bedroom, you turned off the lights and waited for the waves of pain to ebb from the lesions, which you picked at often until they bled and crusted, making you feel ugly, opened up. You always slept facing the wall, afraid they would look at you and leave.

Conclusion: It is possible that the proximate cause of death was moving in together too quickly.

Up close, a whale might appear all baleen, all fluke, all flipper. A whale's whole body only comes into view at a distance. But this perspective flattens it, removing its stellate barnacles and grooved scars and what you can see in its eyes.

Before submarines or underwater photography, most people only ever saw whales up close after the animals had died, stranded on some beach in varying degrees of decomposition. In the seventeenth-century *Historiae naturalis de quadrupetibus libri,* or *A natural history of things that had (or once had) four legs,* the Polish physician John Jonston etched a series of whales as they appeared on land, their fins sunken by gravity. The specimens are all clearly modeled after dead or nearly dead whales—each sports a lolling tongue and a helplessly extruded penis. Though certain etchings are recognizable—the first two appeared to me to be sperm whales and the last perhaps a pilot whale—the others are fantastical. One resembles an alligator-fish chimera, with a corrugated back, clawed fins, and a beaked, toothy smile. It's clear Jonston did not exactly draw whales as much as he drew his idea of the whale, the specific iteration of how decay wrought something that used to be almost superlatively alive. Half-imagined drawings like these created enormous confusion in whale taxonomy, leading naturalists at the time to propose more than a dozen living species of sperm whales (there are three).

The corpse of a whale offers a blueprint of its internal self, the arrangement of its organs and vessels and heart. But a dead whale offered naturalists little insight into how the whale may have lived. That knowledge was reserved for the people who hunted them. They mostly saw live whales in glimpses—a fin, a fluke, a spouting blowhole. But sometimes, before the hunt, whalers would get lucky and spy, somewhere in the distance, the entire body of the whale, its leathery form seizing air like some supernatural bat. But gravity makes these moments, when you can grasp the full nature of a thing, fleeting.

In New York City, in the waters by the Rockaways, it is easy to see whales. I learned this last year on a whale-watching boat called *American Princess*. The trip was nearly four hours, so I bought a hot dog and seltzer (elsewhere on the boat, the real pros smuggled aboard a platter of shrimp cocktail and two bottles of sparkling wine). As we left port, a volunteer naturalist with Gotham Whale, the city's marine mammal advocacy group, narrated the excursion. "There's no guarantee we'll see a whale," she said, adding that this disclaimer had become necessary after an unlucky cruise left some would-be watchers asking for their money back. The *American Princess* did not control the whereabouts of the whales, she explained. When she told us that in the eighteenth century, before they were hunted to near extinction, New York's waters teemed with whales—humpback whales, fin whales, sperm whales, and right whales—people around me gasped, muttered to one another, and looked around the waters as if the long-dead whales had left some sign of their past abundance.

For most of the four hours, we saw things that were not whales. Birds: an American oystercatcher at the pier, a few pelicans, a sheet of seagulls. Garbage: ghost nets, ship hulls, a bobbing can of beetle-green Sprite. Skylines: the topsy-turvy chaos of the Coney Island strip and a distant brown smudge that someone suggested might be New Jersey. We had been informed that the best way to look for whales was to look for the traces they leave, such as splashes and strange ripples in the water. So we kept our eyes glued to the horizon, mistaking every large wave for something more.

A few hours in, the captain shouted—"Whale, four o'clock!"—and we all whirled around in our best guess of an analog watch face. I was almost too late, but I am sure I saw it, a white-crested humpback erupting sideways out of the water. The whale blinked out for a split second, return-ing to the water in an enormous splash. Suddenly the boat was abuzz. Everyone wanted to know who saw it, how big it was, how much of its body touched air. Someone emerged from the bathroom and, upon learning they missed the whole thing, threw their hands up in disbelief. My partner, T, who was on the boat with me and had missed the whale, asked me what it looked like, and I fumbled for a more pre-cise description than "big, gray, and whale shaped." I knew I had seen a whale, so it must have looked like one.

Necropsy Report: A Relationship (*cont.*)

History: Two months into dating M, you came out to your mother. You had known you were queer since June, but you waited because you knew she wouldn't take it well (she didn't) and also because you wanted a partner to point to as proof that this wasn't all in your head. You'd never had a serious boyfriend before, and you wanted your mother to understand you'd found someone who made your amor-phous longing make sense. You showed her a photo of M and the first girl you dated, and she asked why you wanted to date girls who looked like boys, and if this meant you're not actually gay. You didn't know how to answer that either and wondered if this made you less gay.

When you flew back east for your last semester of col-lege, you doubted many things but were sure of one: as long as you stayed together with M, no one could tell you that you weren't queer.

Conclusion: The proximate cause of death may have been your inability to imagine how to be queer and alone.

In February 2016, scientists off the coast of British Columbia tagged a young killer whale named Nigel, or L95, while he was swimming with his family. He was twenty years old and lived in a community of orcas called the Southern Residents. The scientists attached a barbed tag to his dorsal fin, hoping to understand how whales forage in the winter. They observed his behavior over the next few days, noting nothing unusual (his ribs pressed up against his skin, something that had become routine for this population of whales). Before the end of the week, the scientists noted, his tag seemed to have fallen off.

Nigel was found dead in April, and scientists towed him to a nearby village for a necropsy. His body was badly decomposed—black skin flecking like old paint, belly distended and gummy pink—but scientists ID'ed him from two puncture holes in his dorsal fin, where a tag had recently been attached. They x-rayed the fin, revealing that it still held seven of the petal-shaped barbs designed to keep the tag under the skin. The attending pathologist found that Nigel had died when fungi entered his bloodstream through the punctures, worming deep into his lungs, and killing him—a rare outcome of routine tagging.

It is nearly impossible to imagine that a healthy whale in the prime of his life would have died from such an infection, but Nigel was not a healthy whale. He was not just hungry when they tagged him; they tagged him because he was hungry. He belonged to a pod of whales debilitated by disappearing salmon and polluted waters and the roaring noise of vessels. There is a way, the scientists realized, to study something to death. But scientists have to study whale death, to understand how and why we cause it—as we almost always do.

In 2018, a killer whale named Tahlequah, or J35, living off the coast of Seattle, Washington, carried the body of her calf, who lived for just half an hour, over one thousand miles. The calf's death was no surprise. Like Nigel, Tahlequah belonged to the Southern Residents, one of the most imperiled populations of killer whales, which are some of the most contaminated marine mammals in the world. Although male killer whales carry their accumulated load of toxins for their entire lives, females can shed toxins in milk for their calves, poisoned mothers unwittingly poisoning their babies. The newly dead baby had been the first born in three years.

In some researchers' eyes, Tahlequah grieved. She swam

with her calf draped on her snout, its taut body eerily alive and white parts still glowing baby orange, eyespots glimmering in the moonlight. When the calf slipped away, she took six or seven breaths and dove, deeply, to retrieve it. She carried him this way for six days, continuously nudging him away from sinking until he no longer looked like a sleeping calf but a dead one, rigid animal form melting away. Eventually the other killer whales in her pod took turns keeping the dead baby afloat, nosing its body around like a beach volleyball. Tahlequah carried the calf for seventeen days, watching what she loved become unrecognizable to her but seeing no other choice than to continue.

Necropsy Report: A Relationship (*cont.*)

History: Near the end of the whale painting class, you had spent more time at the New Bedford Whaling Museum than seemed humanly possible, and the sight of a whale no longer struck you with that old sense of wonder. In your mind, they had become a commonplace shape: circle, triangle, square, whale.

You and M worked on your final projects for the whale painting class on opposite sides of the California king bed you had inherited from the prior occupants of the apartment. The bed, which had once felt like a luxury, now recalled the physical incarnation of the space widening between you. The air between you seemed to have grown more tense, the mood darkened, but you could not tell if this was all in your head.

Later that week, M introduced you to someone in a coffee shop as a friend, no girl-. You began to spiral. You worried they were ashamed of you. Maybe you were a bad lover, still learning the ropes of queer sex. Maybe you just weren't queer enough for them.

External examination: When they finally broke up with you, it was both anticlimactic and unsurprising. There was not much to say, because, as M suggested, there was never much between you in the first place. The next day they came by your apartment and you handed over their global-warming whale painting, coiled up in a Trader Joe's bag.

Conclusion: The proximate cause of death may be falling in love with the idea of a person, or the idea of a relationship.

In 1987, sonar on a routine submersible survey of barren seafloor in the Santa Catalina Basin picked up on something almost supernaturally large, 4,000 feet below the surface of the ocean. It was a whale skeleton, 65 feet long and sunken in the sand. Without the buoy of inflated lungs, a dead whale makes it to the seafloor relatively untouched, and then the scavengers come. This one had been dead for years, but its remains had become a teeming city in the mud, nourishing clams, mussels, limpets, and snails. This was scientists' first encounter with a whale fall, the most benevolent kind of burial.

The oceanographers, led by Craig Smith of the University of Hawai'i, returned a year later and observed what they found living on the bones, including many species new to science and others that were only known from deep-sea hydrothermal vents. Some of the mollusks found on the whale contained chemosynthetic bacteria, which draw energy from chemicals instead of photosynthesis. Entire ecosystems depend on these deaths, creatures whose lives revolve around chance windfalls of blubber, gut, and bone.

Whale falls linger for decades, feeding scavengers in roughly three stages. The first invites mobile scavengers seeking flesh: sleeper sharks and bludgeon-headed rattails, hagfish and isopods. They swim from afar and congregate to strip the carcass to the bone, with sleeper sharks ripping off soft tissue in chunks and hagfish rasping at the flesh. These scavengers work fast, devouring the equivalent of a small person each day, but still it can take two years to strip all the flesh.

The second stage attracts fewer species of scavengers but far greater numbers of them. Invertebrates reign here, with worms that resemble shag carpets and tiny crustaceans called comma shrimp feeding on the rich organic material left inside the bones and scattered in the surrounding sediment.

The third stage of whale fall, called the sulfophilic stage, happens when the carcass is reduced to a skeleton. The creatures that arrive have specifically evolved to feast on the lipids locked inside the skeleton. Dense, glowing meadows of bacteria descend on the skeleton and feed on the fats and oils inside the bones. Feasting like this produces hydrogen sulfide, which sustains chemosynthetic clams and gutless, mouthless bone-eating worms. The worms root themselves in bones like blossoms and feast on the lipids. They have names like bone-eating snot flower. Their red tufted gills cluster into fringe, adorning the skeletons in rosy tapestries as they soak in oxygen from the water. The small, soft worms have the power to disappear a whale skeleton, scattering larvae into the current that will drift until they encounter bone.

In the largest whales, this third stage can outlive many of us, lingering for as long as a century until all that is left is the mineral husk of bone. It even outlives the whale itself, surpassed only by the bowhead whale, which may live to be two hundred years old. If a whale's life is a marvel, its death is its legacy.

Necropsy Report: A Relationship (*cont.*)

External examination: After the breakup, you moved to Seattle for a job. You did not speak to M for six months yet you still thought about them. They had a new girlfriend, and sometimes you thought about her. You scoured the apps and went on dates with people you hoped would remind you of M, and they never did. All this flaccid yearning felt shamefully maudlin—you knew they were not thinking of you—but you didn't know how to stop.

Conclusion: The proximate cause of death is incompatible desires, and the resultant loss of a sex life.

Conclusion: The proximate cause of death is a loss of desire on one end and self-sabotage on the other.

Conclusion: The proximate cause of death could be anything, and you are beginning to see the futility of this tortuous interrogation.

According to the oceanographers who found the first whale fall, there are 690,000 skeletons of the nine largest whale species decaying on the seafloor at any given time. This is to say: when we killed 360,000 blue whales and hauled their bodies to land, we caused another, unimaginable ripple of death at the bottom of the ocean. Hagfish, octopuses, sea snails, bristle and bone worms, adults and larvae, shuttling themselves along the great expanse of the deep sea and coming up with nothing: no whales, living or dead. One whale provides as much food as a thousand years' worth of marine snow, the white flakes of organisms that died and disintegrated nearer the surface. In the historically whaled waters of the North Atlantic, about one-third of organisms that specialize on whale falls may have already gone extinct.

Necropsy Report: A Relationship (*cont.*)
Conclusion: Slowly, you stop thinking about M.

Conclusion: You go on a slew of bad dates: the bald barista you didn't know was still in college, the iron welder who lives in a barn more than five hours away, the cashier at the Doc Martens store who asks if you want to see how many clothespins they can attach to their face (here, at least, you are intrigued). You feel less intent on finding a girlfriend and more interested in going out to that one queer party, the one that makes you feel like you could fall in love with everyone in the room. You wear suspenders once and immediately regret it. You buy a yearlong subscription to the Crash Pad Series, a queer porn site, and spend your nights watching tutorials on the many ways to have queer sex, the many bodies it can involve, the different kinds of ectasis. For the first time in a long time, you feel like you know yourself.

Conclusion: You learn to be queer and single, but not alone.

Scientists say there can be a fourth stage of whale fall. They call it the reef stage. The bones, sapped dry of all their fat, are reduced to mineral remains. The whale is no longer food but terrain. If the bones are not buried, they become a part of the landscape. The expanse of deep sea is dominated by soft mud and silt, and suspension feeders drift in the deep water in pursuit of hard fixtures like bone, onto which they latch and settle for the rest of their lives. Scientists have observed this final stage in a whale bone spotted in abyssal plains nearly three miles deep, somewhere between Hawai'i and Mexico. The bone had become glazed with manganese, an element that precipitates out of seawater over thousands of years, suggesting the whale sank more than ten thousand years ago. A trio of anemones, clasped on the metallic fossil, unfurled their tentacles like fireworks to catch any food that drifted deep in the abyss. The anemones had found a home on the remains of a creature once so staggeringly alive that it inhaled metric tons of krill each day and fertilized entire food webs with its waste, its hundreds of pounds of heart beating through the water with no sense of what was to come.

Pure Life

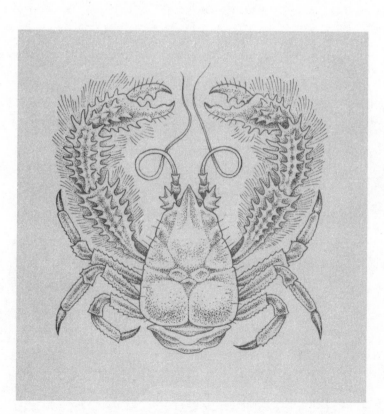

Deep enough below the surface, the ocean's pressure can crush anything unaccustomed to it: a Styrofoam cup, a marshmallow, a human bone. At 100 feet below, the spongy lungs of humans begin to contract. Emperor penguins can dive deeper than 1,500 feet and swim nearly half an hour on a single breath, slowing their hearts to as low as three beats a minute. Humboldt squid descend nearly a mile during the day, only rising to the surface at night. Narwhals can plunge to about a mile, squeezing their rib cages the deeper they go, and Cuvier's beaked whales have been recorded diving just under two miles—the deepest of any marine mammal. At more than 7,000 feet below, where thousands of pounds of force are exerted on every square inch of surface, yeti crabs do just fine.

In 2005, on a hydrothermal vent south of Easter Island, a submersible named *Alvin* sucked up a six-inch crab from the seafloor with a vacuum hose called a slurp gun. At the surface, scientists examined the crab—its carapace the shape of an egg and pale yellow like the moon, with legs resembling feather boas. It was not just a new species but a new family. They named the family Kiwaidae, after Kiwa, a Polynesian deity associated with the ocean, and the crab *Kiwa hirsuta,* after its abundance of body hair. But the crab earned its catchier moniker from its resemblance to the shaggy white legend of Himalayan folklore, though its own hairs are less soft shag and more toothbrush bristle. That original crab now floats in a jar in Paris, on a shelf in the archives of a natural history museum, surrounded by other bottled crustaceans reeled in from the deep.

On the internet, one species of yeti crab, *Kiwa tyleri,* became the subject of a popular meme. In it, a lone yeti crab perches on a craggy rock above the caption: "This creature has adapted to the crushing pressure and oppressive darkness." When I first saw this meme, several days after Donald Trump was elected president, I felt a kinship with the crab. I made the meme my cover photo on Facebook, which felt much more biting at the time than it does now.

When I return to the meme years later, I see the cloudiness of the metaphor. Darkness has no moral value, and its omnipresence matters little to a crab with no functional eyes. And pressure is relative, depending on the body that moves against it; what crushes a human behooves a blob-fish. Though the yeti crab's environment seems inhospitable to us, it is nothing to be pitied. The pressure does not crush the crab, and the darkness does not oppress it. It is exactly suited to the life it leads, however strange or repulsive we might find it. What use is the sun to an eyeless crab? It already has everything it needs.

I moved to Seattle in the fall of 2016, a few months before Trump was elected and the sun left for good. My new neighborhood was more than 80 percent white and yet advertised itself as "the center of the universe." Fremont's county council proclaimed this status in 1994, when the city was a haven for artists. But Fremont would be gentrified in a matter of years, taken over by offices belonging to Google and Sporcle, and other tech companies with insidiously charming nonsense names. Recreational marijuana was legal in Washington, and we lived a few blocks away from a pot shop owned by a white man named after the Buddhist term for enlightenment.

I'd moved to Seattle for what I thought would be a dream internship. But the company was overwhelmingly white; the #poc Slack channel had only two other people, one of whom was my boss. I'd also just started a regimen of a grueling drug called Accutane. It was meant to purge the painful cysts that had erupted from my face since middle school, but the side effects—muscle aches, fatigue, dandruff, depression—made me feel like I was no longer in control of my body.

On the night of the election, I came home to a party that had turned sour. Half-drunk beers littered the floor, and people had stopped talking. I didn't want to be at the party, but I couldn't bear to be alone, so I turned on *Blue Planet*. In the episode, a gray whale watches while a pod of orcas devour her calf. She sees it coming from a mile away but has no way to stop it. The calf unravels into white and red ribbons in the orcas' jaws until she vanishes completely in a frothing, burgundy cloud. The mother lingers but cannot stay, and she sets off again into a sea that has never felt emptier.

Until relatively recently, scientists thought that all life was dependent on the light of the sun. They knew that plants, the anchor of our food chain, spun sugar from sunlight in photosynthesis, and every other living thing ate plants or ate something that ate plants. Our imagination was not vast enough to look beyond the surface, to conceive of another way of living on Earth.

Scientists first discovered hydrothermal vents in 1977—not in search of an alternate way of life but in search of heat. Scientists suspected pockets of warmth existed in the deep sea as early as the 1880s, when a ship named *Vitaz* pulled up samples of water from depths of 2,000 feet that were, strangely enough, warmer than waters at the surface. Scientists assumed these hot spots happened by way of the dazzling power of the equatorial sun, which heated water at the surface until much of it evaporated, leaving swaths of densely salted waters that sank into the deep. But in 1964, the ship *Discovery* pulled up waters deep in the Red Sea that were strangely hot, at 111 degrees Fahrenheit. And the next year the *Atlantis II* pulled up sediment from the bottom of the Red Sea and found it was 133 degrees Fahrenheit: black ooze too hot to touch.

By the 1970s, one of the most pressing questions driving this discovery was science's notion of "missing heat," or the heat that should be escaping Earth's mantle as the rocks radioactively decay. Scientists built probes and wedged them into the seafloor to measure the heat flow by mid-ocean ridges, where the mantle rises toward the surface, but they were puzzled by what they found: waters by these mid-ocean ridges were far cooler than expected. Clive Lister, a scientist from the University of Washington, suspected that deep-sea vents could explain the mystery of the missing heat. If seawater trickles into the porous rocks of oceanic crust, warms from the magma underneath, and then rises again to surge through small vents in the seafloor, this would create heated pockets of water in the deep sea. The process would be just like boiling a pot of water. The heated water rises to the top of the pot, releases heat into the air (or in the vent's case, the water above), and then sinks back to the bottom as cooler, denser water. In 1977, scientists from the Woods Hole Oceanographic Institution set out for the Galápagos Islands to see if such vents existed.

Wintertime in the Pacific Northwest felt subterranean: dark, frozen, eternally wet. And absurdly, in a city known best for its constant rain, hardly any of the bus stops by my house had overhangs, so I was regularly left dripping and downtrodden like some sad Seattle punch line.

I had a few good friends in the city, but many of the young people I met worked for one of the giant tech companies that I vaguely understood to be responsible for sucking the soul out of a city I barely knew. I went to party after party thrown by kids my age who worked in tech. At one, hosted by a college classmate who lived in a luxury apartment and worked for Microsoft, I told everyone there how unhappy I was, and how alone I felt. "There's definitely a ton of colored people here!" a white guy told me, swaying with his Solo cup. "I mean, people of color. That's what I said, right?" I craved intimacy, not just holding someone and being held but the closeness I felt with my queer friends in school and back home. I wanted communities that warmed me until I tingled.

On February 15, 1977, the Woods Hole scientists lowered a steel cage studded with cameras, sensors, and strobe lights—the contraption was named ANGUS, short for

Acoustically Navigated Geophysical Underwater System—to the bottom of the sea, a spot a few hundred miles west of Ecuador and northeast of the Galápagos Islands. ANGUS scanned a little more than seven miles of seafloor, transmitting the temperature of the water to the scientists' ship. For hours, ANGUS signaled a constant temperature of 35.6 degrees Fahrenheit, near freezing. But around midnight, ANGUS passed over an area with an unusual temperature spike. It lasted three minutes before the waters returned to near freezing. When the scientists developed the photos onboard, they saw a martian landscape: lava squeezed out of the seafloor like mounds of toothpaste. Frame after countless frame, the footage showed the same empty lava landscape. Then, when they reached the point in the footage that registered the midnight temperature spike, they saw an explosion of life: white clams and brown mussels littering the lava flow. The mollusks were visible for just thirteen frames—the length of the temperature spike—and the only hot spot of life found in ANGUS's three thousand photos.

On February 17, scientists dove in the deep-sea submersible *Alvin* to see the clams and mussels with their own eyes. They uncovered a patch of ocean shimmering in curtains of heated seawater bubbling up from the lava terrain. Almost as soon as it emerged, the hot water turned a cloudy blue, the sign of manganese and chemicals precipitating out of the water. And there was life, so much of it, a snowfall of milky white clams, some the size of books, across nearly 165 feet of oasis. There was even a purple octopus, seeking prey. Soon after, *Alvin* dove again to new vents in the area, and scientists glimpsed a menagerie of creatures they had never seen before: giant white tubes with heads as red as blood, a bobbing gelatinous orb that looked just like a dandelion, tethered to the seafloor.

No one had expected to find such an abundance of life in waters so deep and so cold. The scientists were confused. Around this time, marine biologists were learning that creatures living in the deep sea could feed on marine snow—the small flecks of flesh and waste endlessly cascading from the surface of the ocean. But here was a dense throng of animals living on rocks, miles away from the sun. How, or what, did these creatures eat? When they opened the first water sample from the vents, the air in the room turned noxious, conjuring the smell of rotten eggs characteristic of hydrogen sulfide.

In the years following, scientists would learn that bacteria and other microorganisms soak in the chemical energy of the vents, sustaining themselves not by the sun but from chemicals inside Earth itself. They create energy by mediating chemical reactions—for example, combining oxygenated seawater and hydrogen sulfide to make simple sugars. The process is fittingly called chemosynthesis and explains how these gushing volcanic cracks in the seafloor can sustain their own kind of life. Just as grass and redwoods evolved to convert sunlight into food, these deep-sea bacteria evolved to convert the energy in a toxic gas to a food of their own.

Hydrothermal vents revolutionized many of science's core ideas about life, how and where it could exist. It is only logical that scientists assumed the strange creatures living on the seafloor would survive on the flecks of fish that died nearer the surface, the scraps of sun-touched society. But these animals eked out an alternative way of life. I prefer to think of it not as a last resort but as a radical act of choosing what nourishes you. As queer people, we get to choose our families. Vent bacteria, tube worms, and yeti crabs just take it one step further. They choose what nourishes them. They turn away from the sun and toward something more elemental, the inner heat and chemistry of Earth.

The yeti crab *Kiwa puravida* dances to live. It makes its home by a methane seep at the bottom of the ocean, where cracks in the seafloor emit gas like a smoke machine. This particular species was first discovered in the seas surrounding Costa Rica, crawling around a mud volcano nicknamed Mound 12. The tip of Mound 12 crests approximately 160 feet above the seafloor, as tall as Paris's Arc de Triomphe. *K. puravida* crabs, clams, and tube worms take refuge on its sloping sides, bathing in the methane and hydrogen sulfide gases burbling out from the seep. But

K. puravida is the only one dancing. The crab waves its enormous claws, each porcupined with prickly setae, above its head. Their rhythm is slow but sure, claws swinging back and forth, shimmering almost like a mirage within the heat of the gases.

When scientists first filmed *K. puravida*'s strange movements, they were struck by this comical dance. There was no other word to describe what the crabs were doing, swaying with such intention, very unlike other crabs. They found the crab's mossy bristles teemed with a garden of chemosynthetic bacteria, tiny organisms that extracted food from the chemicals of the seep. The carbon isotopes and fatty acids of the crab's body revealed its diet consisted mainly of bacteria, rather than anything that made its life by photosynthesis. The scientists realized then that when the crabs danced, they were farming the bacteria clinging to their bristles. Waving their claws back and forth ensured water with fresh oxygen and sulfides would bathe them, nourishing the bacterial meadow. The crabs were farming their own food at the bottom of the seafloor, miles from the sun.

Wouldn't dancing all day and all night make any creature, crustacean or not, tired? But according to researchers, dancing doesn't exhaust the crabs. After all, they wouldn't dance unless it gave them energy.

Two months after I moved to Seattle, an acquaintance from college invited me over for dinner. We'd never been close, but her presence and her space radiated warmth. I rambled about everything and she listened, patiently, before asking me if I'd heard of a monthly party called Night Crush. It was thrown by queer people of color for queer people of color, she said, adding that her friend performed at the party, go-go dancing for tips. She told me it happened the first Saturday of every month, meaning the next day. On the bus ride back home, I scrolled so deep through Night Crush's Instagram that I missed my transfer: photos of mostly people of color dancing in mesh, sequins, strappy thongs, and unitards. The dancers were Black, brown, Asian, mixed, fat, in wheelchairs and out of wheelchairs, all glittering. Scrolling, I could almost feel the sweat.

We showed up embarrassingly early—the bouncer was eating a ham sandwich by the DJ booth and had to be called to stamp us. We walked to the dance floor, a large black box crowned with a glinting disco ball, and watched the DJ spin Rihanna to the empty room, vocals glancing off the walls, from a booth decorated with a banner reading PAY ME, DO NOT FETISHIZE ME. I danced so hard all night that I didn't pee; my sweat made me as moist as a salamander. There were moments when the whole room vibrated together and I could have sworn my feet left the ground, lifted by the bodies swaying and shrieking around me.

The days grew shorter, the afternoons darker, and the ambient fog sharpened into sleet. But Night Crush had become my oasis. I went with my roommates, with unsuccessful Tinder dates, with my ex's best friend and their Tinder date, with anyone who wanted to go. It was always too full, with crowds that lifted you at the crescendos of certain songs. But even at its most packed, a chaos of glitter and pleather and binders and denim and fishnets and lipstick, the crowd never jostled for space. We would part to let the bartender through to collect empties, we made space near the front for people who had money to tip the dancers, and we left enough room for people to dance comfortably, safely. When the heat of our bodies and breath grew overwhelming, we stepped into the dank mist of the street to smoke and shiver together until we were ready to go back in. We understood that this space, for this night, was all we had. We knew it wouldn't return for a month, so we savored it and made it last.

Like any oasis, a hydrothermal vent has edges, and limits. The heat it spews can reach only so far, forcing the creatures whose lives depend on it to gather close, and in swarms. Straying too close to the gushing, volcanic vents themselves could boil a crab alive. But straying too far is a sure route to waters so cold they can paralyze. Scientists long suspected that the lethal consequences of deep polar water explained why so few crustaceans lived in the waters encircling Antarctica.

This margin of safety is slim and precarious for any yeti

crab but especially those in Antarctic waters, like *Kiwa tyleri*. When scientists discovered *K. tyleri,* they found as many as seven hundred of the milky-white crabs, some as large as avocados and others as small as peas, in one square meter. At a distance, the vents seem to be buried under snow. The entire crab population is packed into a few cubic yards where the waters hover at a hospitable 77 degrees Fahrenheit, a ring of shelter between the freezing cold sea and a column of vent water that can exceed 700 degrees Fahrenheit. The crabs don't seem to mind living on top of one another. They share space, scuttling over each other's backs and crowding tightly together until no sliver of rock can be seen, only bulging hills and valleys of white crab shell. Their bodies are squat and compact, making it easier to hold fast to the vertical walls of the vents and allow more crabs to cram together on the seafloor below. Caught between frigid and boiling waters surrounded by wasteland, the crabs have nowhere else to go; they must find a way for this one small safe haven to accommodate all who need it.

On the first Saturday of December—my third Night Crush—I woke up and read the news. A warehouse in Oakland called Ghost Ship had caught fire during a concert. Two of my roommates were from the East Bay, and we scrolled through updates that came infrequently, scanning each new story for the names and faces of anyone we knew. Nothing was confirmed, everything uncertain, and it felt almost wrong to go out. I checked Night Crush's Instagram and saw they'd posted about the fire: "we find ourselves in another situation where no words feel quite like the right words," the post read. "for those missing, for those dead, for those impacted by these great losses. for the folks who find freedom in our night spaces our music spaces our art spaces our dance spaces our club spaces— we are thinking of you." The post continued: "we are gathering tonight and hope to find life and comfort with one another." At some point that evening, getting ready or in the car, we saw that a man we'd gone to school with, who'd gone to the concert, was missing. Night Crush that night felt different, somber and desperate, but we were no less loud. I writhed and sang and looked up from the darkness into the silver scales of the disco ball. I tucked mossy green bills into netted tights and garters. I felt glad to not be alone, but I found myself breathing deep, stepping outside often to remind myself of the way out, to stare into the dark pit of the sky.

The following week, we learned that the man, Nick, had died in the fire. I'd never met him, but I'd seen him perform in his band somewhere in Providence, probably in a warehouse. I read his obituary in my hometown paper and saw that the reporter had interviewed someone I knew, a graduate student in my department named Mark. Mark had long brown hair and wrote strange, short poems every day that I found polarizing and didn't entirely understand, which perhaps was the point. When Nick died, Mark was hitchhiking barefoot across the country to protest climate change. The next month, Mark would be killed by an SUV in Florida on the 101st day of his hike. Four days later, another one of my classmates, Jess, died of brain cancer.

The pain of Nick's, Mark's, and Jess's loved ones is not mine to intimately feel, and the loss they still grieve is something I will never understand. But that winter my world felt destabilized, operating in some alternate timeline in which people I knew were dying too young. It was less than a year after the shooting during Latin night at Pulse, a gay nightclub in Orlando. The president was swiftly and successfully dismantling the few protections in place for queer and trans people. An energy company was installing a toxic, unsafe pipeline on land stolen from the Standing Rock Sioux Tribe. And all the while in Seattle we were reminded of the Big One, an earthquake everyone knew could strike at any time, a magnitude 9.0 in the next ten years. Nowhere felt safe, which meant exit signs felt like a lie.

After *Alvin* discovered the first hydrothermal vent site in 1977, scientists from Woods Hole kept finding more. They named each after the creatures that proliferated off the vents: Clambake 2, Dandelion Patch, Oyster Bed, and Garden of Eden. Rose Garden, discovered in 1979, was named after the vent area's dense thickets of giant tube worms, their white, slender, stalk-like bodies swaying below their blood-filled, feathery gills. When scientists first glimpsed the site, they were struck by how the tube worms resembled long-stemmed roses. The worms' gills course with hemoglobin, capable of transporting oxygen and sulfide at the same time, allowing the worms to create their own energy. In 1985, scientists returned to Rose Garden, expecting to see the same dense forest of tube worms. But the white-stalked, red-lipped bouquets were gone. The site had become overrun with mussels and clams, species clearly able to outcompete the tube worms for food.

In 2002, researchers returned to Rose Garden. But when they dispatched *Alvin* in the same spot in the Galápagos, the site was once again unrecognizable, swallowed up by a black bed of erupting lava, a fresh volcanic eruption. The lush oasis, one of the first glimpses scientists ever had of hydrothermal vent life, was gone.

Scientists now understand that these vents don't last forever. Though some vent fields spew sulfides for thousands of years, individual vents can last only a few decades. These life-giving spumes can vanish at any moment if their heat source is cut off, perhaps undone by an errant earthquake or volcanic rumbling. A whole world extinguished by something uncontrollable.

If being queer in a city grants you the privilege of going to queer bars, queer clubs, and queer parties, it also means grieving them when they disappear. The first lesbian bar I went to, the Lexington Club, closed in 2015 in the wake of rising rents in its historically queer, gentrifying San Francisco neighborhood. It had been open for eighteen years. The Lex was a gloriously divey spot with an electric-blue bathroom blackened by names and messages scrawled in Sharpie. I'd only ever been to the Lex when I thought I was straight, and the bar closed a month before I came out.

On those first Saturdays in Seattle waiting in line to get into Re-bar, the club that hosted Night Crush, you could throw a stone in any direction and find construction. The entire surrounding area had been rezoned from seven to forty stories, and luxury apartments were shooting up in gray turrets, orange and yellow cranes clustering around their bases like licking flames. The owners of Re-bar had spoken publicly about the skyrocketing rent in Capitol Hill, Seattle's gayborhood, which had been diluted in recent years as tech transplants priced out many queer and lower-income long-term residents. As spring came, an influx of white people began showing up to Night Crush, ostensibly friends of those the party was meant for and unaware of how their presence could be seen as intrusive. Rumors started to spread about Re-bar closing, about the building being sold, about it becoming luxury apartments named after yet another musical note.

Few institutions last forever, and bars close all the time. But when a place is your only port, your harbor from the elements, its closure means the loss of something sacred. Of course there will be other bars, other clubs, other parties, but will they value you? Will they prioritize your safety and your joy? Will they protect you from the cold?

I first learned about the work of the artist Sable Elyse Smith after reading an essay by the writer Jenna Wortham, on gay parties and corporate pride. Smith's essay "Ecstatic Resilience," which characterizes the club as a "sanctuary for queer liberation," is visually broken up in stepping stones, an archipelago of paragraphs against cold white space. "This is the dance," Smith writes. "This is the work of living between that distinct moment when we push past exhaustion to elation. Ecstatic resilience glittering down."

A few weeks before I left Seattle, I went to Night Crush for the last time. All night I said goodbye to the city, close to weeping, squeezed by strangers, singing in unison, soaking in our sweat, inhaling air that felt heavy with our collective breath, bathed in the spinning flecks of light from the disco ball. When I left the club, my muscles oozed but my body pulsed from dancing. I felt lucky to be alive.

Since 1977, scientists have discovered hundreds of hydrothermal vents across the world's oceans, along with the many new species of animals that thrive by these curious ecosystems: yeti crabs, new species of jumbo clams, shrimp with eyes that perceive thermal radiation. Each site was home to a community that had engineered their own form of energy.

For many researchers who study the creatures clustered around the vents, the most prominent mystery is their resilience—how, exactly, they manage to persist over time and across disasters. It is no small feat to build a life near an active volcano. It could erupt in an instant and raze a community or burn out slowly, taking its life-giving warmth with it. Researchers do not understand how species of yeti crabs have found safe havens around the world separated by hundreds of thousands of miles of water so cold it slows the human heart. Even setting aside the frigid temperatures, the distance seems untraversable for a sightless crab. But as submersibles around the world have continued to search for vents, they find more and more, some smaller than others and tens of miles apart. One scientist proposed that cold seeps, fissures on the seafloor where chemicals like methane burble out, could act as stepping stones between vents, warm safe houses that could offer the animals safe passage in the deep. In 2006, scientists from Woods Hole sent *Alvin* out once more to study a series of recent eruptions south of Mexico that had wiped out familiar vent communities. They sampled the oceanic currents flowing by these vent sites and found that larvae can colonize vents more than two hundred miles away. But I admit I am attached to the mystery of these spaces and how it keeps them sacred, an impossible, shimmering way of life that we were never meant to understand.

When *Alvin* returned to the once-teeming Rose Garden in 2002 only to confront a cataclysm preserved in cooled lava, the scientists were crushed. But surveys nearby revealed a miniature garden in its own right: tiny tube worms and walnut-sized mussels. They realized life here probably took root in the wake of the lava flow that had quenched Rose Garden. They named the site Rosebud. In 2005, Woods Hole researchers returned to Rosebud to find the communities flourishing—mature creatures shimmering in the chemical warmth. The Rosebud site, scientists wrote, highlighted the dynamic nature of deep-sea sites. Oases here, where so few things are certain, inevitably blink on and off. But life always finds a place to begin anew, and communities in need will always find one another and invent new ways to glitter, together, in the dark.

Beware the Sand Striker

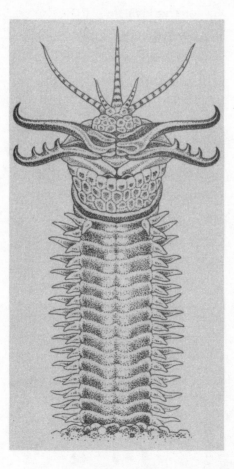

It is unnerving to see a worm that is longer than a man. But this is how long the marine worm *Eunice aphroditois,* also called the sand striker, can grow. The longest specimen on record was nearly 10 feet long, found hidden in a float of a mooring raft in a harbor in Shirahama, Japan. The worm appears in a photo draped across the full length of a crosswalk. A man crouches in the background, on the other end of the street, dwarfed by what appears to be a brown rope.

When seen in certain lights, the sand striker is beautiful. It shields itself in a hard, iridescent exoskeleton that shifts from purplish red to shadowy black. In the light, the worm glitters. Its jaws are magnificent, sprouting from the head like elk antlers with serrated edges. It has antennae too, zebra-striped feelers that twirl upward to sense for prey. For its entire adult life, the worm buries its body in the sludge on the seafloor, invisible to anyone who passes. It only emerges from its burrow to hunt.

When it does, it looks like a blooming. The silt rustles. Its antennae swivel. The worm vaults upward as its powerful jaws flip closed like a bear trap. The sand striker is a carnivore and often feeds on other worms, drifting carcasses, seed shrimp, and mollusks. It also feeds on prey much larger than the span of its jaws, such as a lionfish as long as a dollar. It has no decipherable face, no complex brain, but it can swallow prey like quicksand.

After it strikes, the worm retreats back into the sand, which settles until there is seemingly no trace of its presence. The sand seems as still as ever. If you were a fish, an arrow worm, or a seed shrimp, something small and vulnerable to the sand striker's jaws, you might try to remember where its burrow lies so you could avoid it. But the sea is full of other dangers—larger fish, sharks, even fishing spears—and soon the memory wavers. It fades from a scene that you witnessed to a feeling that creeps up when you swim by that coral, that rock, that dimple in the sand, where something deep within you says, *Get out while you still can.*

One day in middle school, I was walking home from Jamba Juice with a friend when we realized we were being followed. She saw him too. Our conversation grew softer and more sparing as we peeked over our shoulders, seeing the same figure in the distance. He was staring ahead, at us. We agreed in whispers to take a circuitous route home. We feigned ignorance, our words louder and laughter faker. We watched him through a compact mirror as he trailed us for nearly six blocks, turning left, then right, then right again. The streets of our neighborhood were empty save for the three of us.

As we turned the last corner leading up to my house, we broke into a sprint and didn't stop until we reached the green-gray gate in front of my house. We pressed our bodies to the earth, fresh-watered dirt seeping into our jeans, unsure how we would know when it would be safe to speak again. I watched a cloud skim the expanse of the sky before I stood up, peeked out into the street and saw it empty, no man, not even his shadow.

This is how I was taught to perceive threats to my body: men in vans and men in bushes. Strange men, unknown men. Men whose danger was clear. Before I left for college, my mother told me to walk at night with my keys between my fingers, fists balled up like sea urchins. This is how she had felt safe in college, an Asian woman weighing less than 100 pounds, leaving libraries past midnight. I tried it once, walking home from a frat party, and felt ridiculous. My fingers cramped, and I doubted I had it in me to slash my keyed hand at anyone anyway. Besides, I felt safe. There were no strange men here, only boys I went to school with.

*   *   *

There are many ways to be a predator. Some pursue their prey openly, in packs or alone. The hunt can be fast: a pod of killer whales chasing a seal. The hunt can be unending: African wild dogs tailing a gazelle with no particular speed or strength, just patience as they wait for the creature to tire. Others hide in plain sight: mantises indistinguishable from the orchids they climb, snow leopards with mottled coats that blur into rocks. The sand striker is an ambush predator, lying in wait, concealed, until a passing cardinal fish or puffer fish comes close enough to strike. The advantage of this mode of attack is obvious: It is exhausting to pursue animals that are nervous and quick, always on the alert, animals that know they are wanted. It is easier to hunt unseen.

The sand striker has two nubby eyes, though they are not what give the worm its hunting prowess. The worm's five striped antennae offer a much better sense of the world than sight alone. The antennae can also wriggle just like worms, enticing fish. When its pricked and swiveling antennae sense prey, the sand striker shoots out of its burrow and catches the creature with its pharynx, a muscular mouthlike feeding apparatus. Prey clamped in its jaws, the worm propels itself back into the seabed, leaving behind an odd pulse in the sand or a small plume of grit, almost like a burp.

The sand striker recoils from light of any kind, sunbeam

or flashlight. When BBC camera crews for *Blue Planet II* tried to film the worms hunting, the worms noticed the artificial lights and refused to leave their burrows. The camera crew could only film them using infrared light, which was invisible to the worms and the people. The only person who could see the worm shoot out of its burrow was Hugh, the camera operator. Sarah, the sequence director, sat on the seafloor in absolute darkness. She must have known that somewhere around her the worm was on the hunt, and might have been able to sense her, but that she would never be able to see it.

There is a story I used to tell at parties, about the first time I gave a blow job. It goes like this. One of my friends, who goes to school in the city and is therefore my coolest friend, knows a girl who is having a party in the woods. She tells me and another friend that she can bring us, and we are elated—our first real high school party. We drive up to the woods in one of our parents' minivan, a hulking thing teeming with blankets, snacks, and blue liqueur we bought at Safeway with a Colorado fake. In the car, we talk about boys, how much we want to make out, savoring those words, "make out." This is all I imagined hooking up to be, making out in cars and on bleachers and in school hallways.

We get there, to a shrouded clearing in redwoods with picnic tables. We drink. We're all seventeen, and none of us is good at it, mixing whiskey with Sprite, shots of Fireball because it's the only thing we can get down straight. One of us starts talking with a boy we don't know and soon disappears into the woods. My other friend and I laugh: she's making out, and we're proud; we know that's what she wanted to do. Without her, we don't know anyone, so we stand around the firepit and my friend takes half an edible, says it's not working and takes the other half. I can't find my drink, and a boy I don't know asks if he can make me another one, and I say yes. "It's strong," he says, handing it to me, and I tell him I'm cool with it, that's what I wanted, and in a way it is. We are teenagers. We are impervious, we think. All of a sudden I realize my friend is gone; she's locked herself in the minivan, painfully high, with all of our food and sleeping bags, and for a second I panic. But I am too drunk to think logically, and so I go back to the firepit and the boy takes my hand and leads me into the woods and I laugh, feeling like I've been chosen. I'm also going to make out with a boy, I think, and we do, for a second, until I think I'm falling, but I'm actually just being pushed down, his hand on my head, tangled in my hair, and his dick is in my mouth, and the first thought that crosses my mind is: *What if I'm not good enough?* I've never done this, but I try, I really try, I've always been a people pleaser, and he moves my head with his hands and I don't remember if he comes but I remember that I stumble and he lets me fall, and sometime later he walks me back to the firepit, holding my hand, and people smile at us when we return, winking at me, at him, and I wink too. I refuse to be embarrassed at this, my first real party. I feel cold suddenly, and I realize he took off my sweatshirt—a green pullover, from a college I will not attend, that is somewhere in the woods—and I am too embarrassed to look for it.

At this point in the story I would pause, waiting for people's reactions, and say: "Wait. There's more." Because when I come back to the firepit, my friends are still gone, and I am alone, thinking about the thing I have just done— *My first blow job,* I say in my head, repeating the words to hear how they sound, how adult I am—and I excuse myself to go for a walk and stare into the sky. The night is crisp, the moon is enormous, and I think I am alone, but a man has followed me. He tells me that he heard I have nowhere to sleep, that my friend has locked herself in the van, and he says he will lend me his sleeping bag, that he is okay to sleep in blankets in his hatchback. "It will be much warmer than sleeping out here," he says, and I relent without thinking, so cold without my sweatshirt. We climb into his car and he reaches across and grabs my breast, so abruptly that I almost laugh, looking at his hand planted on my shirt like an extra appendage.

The rest is so predictable: him grabbing more and more of me, me saying no, or, rather, no thank you, him smiling and laughing and me figuring things must be okay, this must be a normal back-and-forth, me taking his cue and smiling and laughing too until my shirt and pants are off and he is pressing his dick, the second one I've ever seen, a real zero to sixty, into the space between my underwear and my thigh, me blurting out I have a tampon in, that I am bleeding, that I will bleed all over his trunk, and him telling me that the least I can do for him is put his penis in my mouth, he is letting me sleep in his trunk, after all, and he has given me his sleeping bag, which means that he will be cold, me feeling apologetic and almost relieved, that we have found a compromise, and as he is pushing my head down, the first boy I kissed that night shines a flashlight into the trunk and asks if we have seen his graphing calculator, because someone has stolen it.

When I told the story at parties, this was always the punch line, the two boys whose dicks I touched staring at each other through the glass window of the hatchback, yelling over a graphing calculator. It was a TI-89, the most expensive one, the kind you could text with and play games on during class. I would have gone looking for it too. The moral of the story, I figured, was that life is absurd, that nothing is ever as romantic as we hope it will be. I told this story because for a long time it seemed the most exciting thing that had happened to me, and I wanted to have an exciting life, and if I convinced myself it was a fun night, then eventually I would believe it.

After he took his dick out of my mouth, the man went to

sleep quickly. I call him a man because the next day I learned that he was thirty. I stayed up next to him in the trunk of his car, wondering what would happen if a person had sex with a tampon in, if the soft cotton would lodge itself somewhere out of reach and puncture something deep within a body. When I pulled out my tampon the following morning, I wanted to thank it.

When I wrote this story down for the first time, a decade later, I narrated it as drily and with as much minute detail as minutes from a PTA meeting: this happened, then this, then this. I wrote it like this because I was trying to see my own experiences as if they had happened to someone else. I imagined this would help me judge, objectively, whether what happened to me was enough to count as assault, whether it was my fault or someone else's. This is something I've been wondering, because I don't tell this story at parties anymore.

For a long time the sand striker was known by another, more infamous common name. "The bobbit, a giant carnivorous worm with jaws as sharp as daggers," intones David Attenborough in *Blue Planet II* before footage of the worm vaulting out of the sand to devour a fish. The nickname "bobbit" refers to an event, an act of violence in a series of violent acts that made up a relationship. On June 23, 1993, Lorena Bobbitt (née Gallo) cut off her abusive husband's penis with a Ginsu knife. Later that year Terry

Gosliner, the nation's foremost expert on sea slugs, was compiling his latest book of taxonomy, *Coral Reef Animals of the Indo-Pacific*. He had a thousand new species to name, including one menacing, predatory worm that could emerge from the seafloor and snap up a fish. When he thought of the worm's phallic shape and scissors-like jaws, only one name came to mind: Bobbitt. This nickname led to a myth that female bobbits cut off the genitalia of their male mates, easily debunked by the fact that the worms have no external sex organs. But the nickname stuck.

Lorena was an immigrant. She was born in Ecuador in 1969, grew up in Venezuela, and came to the US on a student visa. She worked as a manicurist in Virginia, near Quantico, where she met John Wayne Bobbitt. Lorena told a reporter she found John handsome, a Marine with blue eyes, a beloved symbol of the country where she chose to make a life. Lorena and John married in 1989, two years after she moved to the US. Their wedding photos are grainy and charming, John in uniform and Lorena with pink flowers pinned in her hair. Lorena said that's when John began abusing her physically, beating her, raping her, forcing her to have an abortion. She says he threatened to have her deported.

In 1991, John was discharged from the Marines, leaving Lorena the sole breadwinner. The lender foreclosed on their house and they moved into an apartment. They broke up and got back together. They'd already agreed to separate in June 1993 when John came home drunk and Lorena says he raped her. He fell asleep. She went to the kitchen, drank a glass of water, saw a knife on the counter, went back to the bedroom, cut off his penis, took it with her to the car, and started driving. She turned off Route 28 and threw the penis out the window into a gravelly field where police would later recover it, put it on ice, and deliver it to the plastic surgeon and urologist who reattached it to John.

For the next year, joking headlines abounded—"The Cut Felt Round the World," "A Night to Dismember," "A Slice of Wife." Lorena, who said she was still learning English, gave a statement without a translator. The morning of the incident, the first person to interview Lorena said she told him, quite directly, that John had raped her. The trial happened while John and Lorena were still married. Vendors outside sold chocolate penises and hawked a wagon of T-shirts reading MANASSAS, VIRGINIA: A CUT ABOVE THE REST above a bloody knife. The T-shirt vendors would make twenty thousand dollars. The Oscar Mayer Wienermobile passed out cocktail dogs outside the courthouse, and a folk group played covers on the theme of the trial—"50 Ways to Cleave Your Lover." During the trial, Lorena's legal team would learn that a Marine Corps family advocacy review committee determined in 1990 that John was abusing her, but she was never informed. John was found not guilty of marital sexual assault (at the time in Virginia, a spouse could only be charged with rape if the couple was living apart or the victim was seriously injured). Three months later, Lorena was found not guilty of malicious wounding due to temporary insanity. They divorced in 1995.

After the trial, John starred in several porn films. He went on a forty-city tour, going on *The Howard Stern Show* to talk about his penis. He got a penile augmentation, which horrified the plastic surgeon and urologist who had performed a miracle of a surgery to reattach it in the first place (he later had to reduce it). He was later convicted in two cases of battering his fiancée Kristina Elliott and then again of abusing his third wife, Joanna Ferrell. He devoted his days to searching for an apocryphal treasure chest buried in the Rockies by an eccentric millionaire. He loves Trump.

Lorena became a citizen the summer after the trial, and her family looked on, proud of her and hopeful that they too could one day become citizens. She re-enrolled in community college and met a man, David, with whom she raised a daughter. She started an organization that focuses on the prevention of domestic violence. She volunteers at women's shelters, speaks at schools. Still, after so many years, she remembers the jokes, the laughing.

Though John did all he could to milk the spotlight, Lorena was the one the media fixated on, the one who became a punch line, an *SNL* sketch, a cheeky aside in a late-night monologue. Lorena, the woman who cut off her husband's penis. Lorena, the butt of the joke.

In 2019, oceanographer Kim Martini saw the docuseries *Lorena,* which offered a major reevaluation of Gallo's story, and immediately made the connection between the woman and the worm. She addressed the marine science community in a post on the blog *Deep Sea News,* asking scientists to reconsider the name. "Bobbitt is the last name of a rapist and domestic abuser that should not be immortalized anywhere," she wrote. She asked that the worm be referred to as the "sand striker," one of the worm's more obscure nicknames. Worm scientists were relieved for another reason, hoping the new name would finally put to rest the many myths of the bobbit worm, which neither has nor slices penises.

I wonder if Lorena watched *Blue Planet II,* if she hoped to be transported by a documentary about the wonders of the ocean only to be needlessly retraumatized, returned to the time she was humiliated and demonized by an uncaring public while her husband was paid to star in a pornographic movie called *John Wayne Bobbitt Uncut.* I wonder when she realized that after the media and the justice system stole her story, science stole her (married) name.

Lorena still lives in Virginia, in a new house, two stories, just a twenty-minute drive from the town she lived in with John. When I learned this, I felt confused about why she never left, surrounded by the ghosts of the abuse and the trial and the hounding by the press. But I also understand the security that comes when you know a place and its ghosts. When you have seen the worst of it and survived.

In college, I began blacking out and hooking up. My friends and I had not been cool in high school, so we went to every party like it was our first. We took shots and drank jungle juice out of buckets. Many of us blacked out, often, to the point where it became a game to piece the night together the next morning in the cafeteria, trying to remember what we had done. I knew vaguely that this happened to me more frequently than the others, but I brushed this off as a quirk, something that made me fun.

I learned not to visibly freak out when I woke up in an unfamiliar room, in bed with a strange man. Most of the time, at the sight of him, some memories would bleed back into my consciousness—his wool beanie up on my head, a yellow quilt, the bristling leaves of a hedge. And he would wake up and see me, and we would smile, and sometimes we would have sex. Sometimes we wouldn't, and I would leave wondering if we had the night before, sticking my fingers inside my vagina to feel for any sensitivity, inspecting my body for marks. These were the best possible scenarios, and sometimes I would text these men back, ask to see them again. I would hope that I had sounded kind, smart, and funny the night before, even though I knew that was a very unlikely character for me to adopt in a blackout. Most of the time, things worked out okay.

Other times, they did not. Once, I emerged from a blackout to see a man sitting on my stomach, slapping me. "Call me master," he said, grunting, and, pinned, I did. When he was done moving inside me, finally asleep, I rolled over and stared at our bodies in the vertical shaft of my mirror. I tried to imagine us walking into my room, closing the door, kissing. I remembered spotting him the night before in a sweaty basement of a house party, remembered how I drooled over him, remembered how beautiful I felt when he picked me up and kissed me on the dance floor. This was what I wanted, I reminded myself, and made myself stay in bed until he woke up and I pretended to wake up too.

Another time, swaying with my friends at a party, I learned I'd slept with someone because his friends were talking about me, loud enough that they knew I could hear. "He raw-dogged that girl on someone's lawn," one of his friends said, locking eyes with me, that girl, and his friends laughed.

Yet another time, I woke up next to a friend of a friend, someone I saw at house parties and group dinners. I remembered him flirting the night before in the kitchen, remembered turning him down. I wondered what had changed, but brushed it off and asked him to leave. How was it! my friend texted me, asking about what I presume was a hookup witnessed by many, if not me. Fun! I texted back, trying to end the conversation. And then it happened again; and one last time; and each morning after, I remembered how he had watched me from across the room until I started to slur and stumble, his blurry face hovering in my peripheral vision, insisting on walking me home. Each time I gently shook him awake, told him that I was having trouble sleeping next to another body, and asked him to leave. And each time I apologized for asking him to leave. Each time he left, I went back to my twin extra-long and writhed, soundless, in shame. And then I took a shower.

I never told these men about my blackouts because it seemed a mortifying thing to admit: that I was a child, too inexperienced to handle liquor, too out of control to be an advocate for my own body. I would go out of my way to share these stories with friends, to get ahead of the narrative. I let them fill me in on the events of the night, nodding at things I could not recall until I had reassembled the night in my mind, as refracted through them. If they were happy for me, then I could be happy for me. The more I bragged about those nights, the more inconsequential they became.

When I first came out to my mother, she asked me if I thought I was a lesbian because so many men had been cruel to me. I knew that when she said "cruel" she meant boyfriends who had broken up with me, but instead I thought of these other men, and for a moment I wondered if she had a point.

Sand strikers, or worms like them, have been around for hundreds of millions of years. Unlike dinosaur bones or ammonite shells, the squishier worms did not fossilize easily. Scientists only know of ancient sand strikers from the very few geological traces they left behind: hard jaw parts, tracks in the mud, papier-mâché-like molds of spaces their bodies inhabited before dissolving.

The oldest we know of lived four hundred million years ago in the Devonian age and grew to around 3 feet long, smaller than today's largest sand strikers but still abnormally giant for its day. Scientists described this species of extinct worm solely based on its jaws, as no vestige of its soft body remains. Another prehistoric sand striker from just twenty million years ago left a series of L-shaped burrows found in sandstone in northeast Taiwan. The edges of the burrows were rich in iron, indicating that they may have been lined with a mucus layer to maintain their shape, and when scientists looked closer, they found feather-like impressions in the rock, suggesting the repeated disturbance of sediment from a lunging creature, striking and retreating, striking and retreating.

These ancient worms may have been different, shorter and softer, perhaps, than modern sand strikers, but they hunted in the same way; ancient fish, too, watched for dimples in the seafloor and antennae swiveling in the grit.

Joana Zanol, perhaps the world's leading expert on sand strikers, told the BBC she had never seen one alive in the wild, only in museums. She traveled to East Timor on a research grant to study the worms and could not find a single one. Zanol knew the worms were there, lurking in the sand, maddeningly close to her and actively affecting the ecosystem, even if she could not take a photo or see them with her own eyes. She knew but couldn't prove it.

A few days after I graduated college, freshly moved into an anchovy-sized bedroom in Brooklyn, I opened Twitter to see everyone sharing the same BuzzFeed story. The art was everywhere, spidery black words laced across a bright-red background, staining my feed. It was the victim impact statement read by Chanel Miller, a young woman who was sexually assaulted behind a dumpster at Stanford. I'd been following the case vaguely, noting how it had happened near my hometown, how the media had fixated on Brock Turner's swim times, but there were so many cases like this, meaning stories of girls touched at parties, in dorms, touched outside their clothes or underneath, stories that usually did not result in police reports. Miller was still anonymous when the statement was published, and I would only later learn our similarities. How we were both half-Asian, how we grew up in the Bay Area and once took art classes at the Rhode Island School of Design that made us feel, among other things, inadequate.

"You don't know me, but you've been inside me, and that's why we're here today." Reading that first sentence, my breath caught. Part of the statement is an italicized list of questions the defense attorney had asked Miller. *Did you drink in college? You said you were a party animal? How many times did you black out?* I stared at these questions, so familiar from my own self-interrogations.

The first few times I revised this essay, more encounters flickered back into memory. A tagged picture on social media, an old story recounted at a party, a name that sounded too familiar on LinkedIn. It's absurd how many men I've slept with have later requested me on LinkedIn. I wonder if this is because LinkedIn is the easiest portal with which to find me. I wonder if it's because my name appears occasionally on the internet, slotted under stories I've written. I always wonder what they want from me—forgiveness, my body, or to connect them with an editor of some magazine.

After reading Miller's impact statement, I opened a browser on my phone, switched to incognito mode, and Googled: "can you consent while blacked out but awake." I scrolled through articles, Reddit threads, and PDFs. Everything seemed to say no, you cannot consent while incapacitated, so I Googled more. I rephrased my queries until they morphed from search terms to incoherent personal questions I knew the internet could not answer: "how drunk to consent." "what if you say yes when drunk but cannot remember." "said yes but don't remember is it sexual assault." "how do you know what you wanted while blacked out." "why do I black out so often."

I wasn't even hoping for a particular answer; I simply wanted someone to plot my experience on a grid, to tell me if it was valid to feel this way or if I just needed to get over it. I read about the clinical psychology professor Kim Fromme, who testified in Miller's case and dozens of others. Fromme has made a living by testifying in criminal cases that a person can theoretically consent to sex while blacked out. Her arguments sometimes convinced me that I was at fault. I found a slideshow about alcohol and consent made by two white men who worked at a risk-management consulting group. "The difficult case," the presentation reads, "is someone who has a high tolerance for alcohol but doesn't display the traditional symptoms due to their tolerance level." That was me, the difficult case.

All nature documentaries share a familiar kind of dramatic irony. When the narrator, whether he is David Attenborough or just sounds like him, introduces you to something small, soft, and witless, you know it will be devoured. This is inevitable. When you see the writhing, sequined mass of a bait ball, meaning a school of fish that swims tightly in a globe, you know they will be picked off by whatever larger thing has caused them to swim so close together.

In *Blue Planet II,* Attenborough introduces the sand striker as the sun disappears, giving way to the electric indigo of night undersea. We do not see the worm for over a minute, instead following the looping path of an ill-fated reef fish. It is hard to watch something and know it will die, but there is nothing I can change. The sand striker lunges into an attack, mandibles slicing through the brine to clamp down on the fish and drag it beneath the sand. Even if the fish managed to escape, there may be other worms lurking nearby. There are only so many places to go on the seafloor.

Why is prey in nature videos always "unsuspecting"? "Stealthy Crocodile Captures Unsuspecting Prey." "World's Biggest Spider Gobbles Down an Unsuspecting Lizard." "Cuttlefish Hypnotizes Unsuspecting Crab." When we watch these videos, we are supposed to marvel at the attack, whether achieved by speed or cunning or brute force. The prey is rarely the true subject of the segment. We see a hare bounding through the snow not to understand how it forages but so that we can see how the Arctic fox ambushes prey. We see sea lions jetting under ice to see how orcas hunt in packs. We see bee-eater birds zipping through the sky to witness the dive-bombing techniques of a bald eagle.

There are exceptions, of course, and this isn't to say that we are told nothing about the inner lives of the prey. We learn about the hare's thick winter coat and shortened ears. We see the seals jump in and out of the water to avoid the gnashing jaws of an orca. We see the intricate community dynamics of turquoise bee-eaters jostling over nesting sites embedded in a cliff. But more often than not the documentary shifts our attention back toward the predator, and ends the segment with predation as climax—hare in fox jaw, seal blood in water, jewel-colored bird in the yellow talons of an eagle.

Though prey can be caught off guard, can be surprised, can even be ambushed, prey is never truly unsuspecting. It has evolved the blueprint of its body in response to, or in anticipation of, trauma. The Arctic hare is blue gray in the summer and white in the winter so it will not be seen. Certain creatures have even adapted to give up a part of their body until it can regenerate: sea slugs shedding papillae, crabs sacrificing a claw, geckos shedding a still-twitching tail as a decoy while they escape. Snakes play dead, butterflies disguise themselves as leaves, octopuses squirt ink. These adaptations are remarkable, and make these creatures exceptional in our eyes, and yet would not be necessary without the constant threat of the predator.

I acknowledge this metaphor of predation is cheap. I don't fault the sand striker for hunger, or for hunting. It works much harder than I do, someone who buys meat already dead and plucked. Part of the reason I find its body gruesome may be a hardwired instinct in the animal in me, an animal that fears snakes and creatures that move like them. When the sand striker snatches a fish and begins to feast, it is not thinking of what the fish is feeling. It has no complex brain and no sense of morality, which means its intentions are never cruel. A worm cannot shirk a duty it does not know. But we can.

I am not writing this to blame the men who have touched me when I was not aware enough to consent. Instead, I hope to place them, like pushpins, on a board of encounters that society has framed as acceptable. I do not know what I was like in these states, what I said, how I slurred. For much of my life, the idea of conflict scared me so much that I would do almost anything to avoid causing a scene. My priority was my pride, not my body. I do not know what I would rather believe: if these men thought I wanted it, if they knew I wasn't there to consent, if they suspected at all and buried those concerns, if they didn't care.

I do not know if these men knew I was gone (the clinical term is "incapacitated," but the only way I can understand it is "gone"), burrowed out of my body for the night. I know, legally, logically, that if I was incapacitated (gone), then I could not have consented. I know that to believe this requires overturning a part of my past that I told myself was fine, pushed to the shaded areas of my mind, and allowing myself to feel. This feels like an awful amount of space to take up. In truth, I am fine, most of the time.

If I stretched out my memory of my life like a ribbon and held it up to the light, whole years would be threadbare: worn patches, rips, holes. In a way, this makes me feel relieved. Whatever happened in those hours of my life is lost to me forever, or, if it still exists, molded into something like instinct. These are the missing hours of my life, time I have lived unconsciously, existing as a physical body in space without the power to understand what I was doing or what was happening to me.

Several years ago a boy I knew in college wrote to me on Facebook to apologize for what he did that one night. I stared at the message, racking my brain for what it could have been. I came up with nothing. I almost texted a friend to ask, but afraid my ignorance of that night would give away my blackout, I did not. I wanted to ignore the message, to block him, but I knew he lived in a city I sometimes visited, knew that some of my friends were his friends. I worried what would happen if he confronted me, if he told someone we knew about my blocking him. So I messaged back—within the hour to ensure nothing would seem weird, so that I could seem impeccably unfazed: "No worries at all!" Seconds later, unsure if my first message implied that something did in fact happen, I messaged again: "I'm sure it was nothing." If I were a more ruthless detective of my own life, more sure that I could love myself knowing all the things I've done and the things done to me while I was not there, perhaps I would have had the courage to ask him what he was talking about. But I am not, so I did not.

No, I am not writing to blame these men, but I also am not excusing them by casting their behavior as something instilled in them by systems beyond their control. Almost every system we exist in is cruel, and it is our job to hold ourselves accountable to a moral center separate from the arbitrary ganglion of laws that, so often, get things wrong. This is the work we inherit as creatures with a complex brain, which comes with inexplicable joys, like love and sex and making out in cars, but also the duty of empathy, of understanding what it means when someone is stumbling.

My experiences are not exceptional, in either their recurrence or their severity. But I want to imagine a world in which the men around me when I was younger could have acted as a safety net, could have seen a drunk girl stumbling on a sidewalk as a person, not an opportunity. I wish they could have seen me and alerted my friends, walked me home without touching me, or even just left me alone. Yes, I was fine, not in danger of choking on my vomit or passing out and hitting my head. I was fine until they found me, and then I wasn't.

In certain waters, sand strikers prey on small fish called monocle breams. The breams are the color of armor, silvery bodies with one dark stripe—unabashedly plain among the flamboyant menagerie of tropical fish. They are the kind of fish, it would seem, who would only make an appearance in a nature documentary if they were about to be eaten. Breams feed on tiny shelled things that also lurk in the sand: copepods, shrimp, and microcrustaceans. They cluster in social groups, aware of the power that comes in numbers, in multiple pairs of eyes watching the horizon for threats.

Though the breams may see the expanse of black sand below them as empty, the worms lurk underneath in their mucus-lined burrows, antennae feeling for flesh the worms can grab. The deck, it seems, is stacked against the breams: the food they seek lies buried in sediment that could easily cloak the worms that seek to devour them.

While observing young monocle breams in waters off Indonesia, scientists noticed the fish exhibit rather unusual behavior. It always starts with one bream, who, eyes pointing downward, notices a suspicious dimple in the sand. Maybe there's a crater, or stray antennae peeking above the grit. The fish inches closer in slow spurts, pausing often to assess any new movement. The others, noticing this fish's strange behavior, follow suit, hovering behind, eyes fixed on the sand. Then the first fish starts to blow, spitting jets of water toward the crater, whirling up sand and revealing the worm that lies hidden underneath. Breams are incapable of harming the sand striker, as delicate fins and even the most forceful water jet can't pierce an exoskeleton. But what they can do is expose and warn. Their spurting alerts others nearby to the worm's stealthy presence in a kind of effervescent whisper network.

Once exposed, the worm might retract just one antenna; other times, it slinks back entirely into its burrow, buried too deep to hunt.

Sometimes the breams only knew to spit after one of their own was taken, snatched and dragged deep into the sand. In one case, other species of small and vulnerable fish, wrasses and blennies, joined the crowd, eyes pointed toward the burrow, memorizing the threat.

The scientists marveled at the breams' collective action. Approaching and mobbing a predator such as a sand striker invites real danger—losing a fin, a patch of scales, even dying. But the scientists never saw the worm fighting back against these mobs. They never saw a fish placed in danger by alerting others to a threat. The breams swim around the reef to forage but never veer far from their home range, refusing to be forced out. They will do what they can to make this dangerous place safe for one another.

When the scientists published the paper on the monocle breams, they called their behavior "novel." And it was, in the sense that humans had never observed it before. But had we even thought to look?

A year after I read Miller's victim impact statement, while working my first real job in media, my friend emailed me screenshots of "Shitty Media Men," Moira Donegan's crowd-sourced Rolodex of men who had allegedly harassed or assaulted people. I scrolled through the list and saw the name of one man who worked for my company, who frequently collaborated with my team, accused of unwanted groping. I rolled the hard consonants of his name around in my head during my commute, thinking about him and his enviable job. I saw him in the cafeteria sometimes, or stood beside him in an elevator. A few years later, someone invited me to speak on a panel he was hosting, and I declined.

Back then, my friends and I went every weekend to a tiny club in Bushwick, with a checkerboard floor and three chaotic bathrooms. We always got there early, before there was a cover, and we were often the first bodies on the dance floor. The fog emanating from the DJ booth was always thick but never enough to obscure the message painted on the back wall of the club in white paint: IF YOU TOUCH A WOMAN AGAINST HER WILL IN THIS ESTABLISH-MENT, WE WILL LITERALLY RUIN YOUR LIFE. Even as the club began to fill, as we began to sweat, we could glimpse the white letters between bobbing heads. When dawn broke and the crowd dissipated, the sign hovered back into focus. Each night it outlasted us.

The night I met my first girlfriend, we spent almost three hours talking in my room. They sat on an inflatable couch, me on the edge of my bed. I was so nervous my words tumbled out of me like marbles. I am sure what I said was incomprehensible, mistaking oversharing for flirting, but I marveled at the space they held for me, and how they listened. When we fell silent, they asked if they could sit on the bed. I said, "Yes." They asked if they could touch my thigh, and I said, "Yes," watching as their fingers circled my knee. They asked if they could kiss me. "Yes." They asked if they could take off my shirt. "Yes" slipped out, and unable to catch myself, I laughed. They asked me why I was laughing, and I told them no one had ever asked me these things before. I remember thinking it was almost silly how careful they were being. Of course I wanted to kiss them, to have them take off my shirt, to fuck. We would eventually break up, of course, a protracted and messy split, and I thought about them like I crushed on them—for a mortifying amount of time. But this memory remains as clear as a gem, this sweet negotiation: Can I? *Yes, yes, yes.*

Hybrids

This essay will not end with me folding dumplings.
It will not end with me eating dim sum with my Chinese grandparents and my white dad, ruminating on how a family like ours wouldn't have been acceptable a century ago. How I, the ostensible protagonist of this essay, may not have existed a century ago.

This essay will not include a scene in which I, a part-time intern at a magazine in New York, was confused with the other part-time intern, who also happened to be half-Asian and half-white, whose name also began with *S*. Nor will it present a scene from that same summer at my other part-time internship when I was informed by a manager that I "don't look that chinky," which I assume he considered a compliment. The essay will not include this scene as evidence that I have earned the right to complain about racism because I, too, have experienced it.

Although this essay will include none of these things, it once did. All these endings and openings and in-betweens are scenes I wrote and then deleted. The problem was not that the scenes were corny—I am corny—but that they did not feel particularly true. Actually, they felt canned. The scenes, as I had written them, regurgitated back at me a tidy acceptance of myself and my identity that I wasn't even sure if I agreed with. A final moment of belonging to assuage my years of biracial unbelonging, and written for whom, exactly?

In high school, if I had read an essay that ended with me folding dumplings, eating dim sum, experiencing my biracial quota of microaggressions, transfixed by my deliriously unreadable face in a mirror, etc., etc., I would have loved it. In high school and college, I gorged myself on these kinds of essays—the first entries in an empty archive that we and many others had been shut out of. I had become so acclimatized to reading about the personal lives and observations of white men named David that I felt jolts of recognition each time the essayist gestured toward a mixed-race experience we shared. This newfound visibility, this representation, swaddled me.

So I wrote one of these essays in college, for the publication where someone had confused me with the other half-Asian intern. I wrote about the experience of being called exotic, which I deemed a well-intentioned microaggression with an easy solution: anyone who wanted to know my race should simply ask, and I would happily tell them. For this, I was paid fifty dollars. I came up with this solution because, I had learned, if I wanted to write about a problem, I had to include a solution. I don't remember if I actually agreed with this solution, but I do know that I was nineteen years old and had no coherent politics of my own. I remember feeling queasy as soon as it was published. *Did I really just offer an open invitation to random strangers to demand the fractions of my race?* Looking back, I realize I had written the essay not just for a white editor but also for a white audience. Like a dutiful little trash compactor, I had digested my messy heap of an identity into a manageable lesson for people who were not like me. I had never considered what a mixed-Asian essay that I wrote for other mixed-Asian people might look like. Or, rather, what a mixed-white essay that I wrote for other mixed-white people might look like. And when I tried to think of one, I was afraid I didn't quite have anything new to say, so I decided it was easier to write about anything, everything, but my race.

And yet—this essay is a spoiler in itself—I have never stopped thinking about my mixed race. My race, or rather my preoccupation with what it means and how I should feel about it, is something that may rankle me for the rest of my life. For a long time, mostly in college, I thought I could resolve this irritation by categorizing myself, meaning obsessing over the particulars of my identity—what I could call myself, what spaces I could occupy, when I could blame other people and when I had to blame myself. I don't fault myself for this obsession; if a person is asked "What are you?" often enough, they may become hell-bent on finding an answer.

In the critic Sianne Ngai's terminology, my own irritation with myself is my ugliest feeling. It is the scab I will always pick at and never allow to heal, my smallest festering wound. In her book *Ugly Feelings,* Ngai cites Aristotle, who defined irritation like this: "The people we call irritable are those who are irritated by the wrong things, more severely and for longer than is right." Ngai's touchstone for irritation is Helga Crane, the biracial protagonist of Nella Larsen's 1928 novel *Quicksand.* Helga's racial anxiety emerges in her unusual, unending irritations. Helga is irritated by the smell of stale food, tarnished silver, and ugly teacups. Her irritations are so outsized and misplaced, Ngai writes, that the reader becomes irritated too, not alongside Helga but at her. I worry this essay will resemble one of Helga's irritations, in which I fixate on something so vain and minor it hardly warrants a whole essay. But what is the other option? Becoming Helga, rattling monstrously at teacups for the rest of my life.

I'm not interested in writing toward some resolution of belonging. Maybe it's a side effect of coming out twice in adulthood, but I do not want to feel resolved about myself.

My experience as a mixed-race person is not fixed but always oscillating, between Chinese and white, longing and irritation, pride and guilt. I want to imagine my mixed-race existence in the present and into the future. I want to think about my mixed-race being—not as a noun but a gerund. I want to imagine how I am continuing to live.

Lately I have been fixated on a butterflyfish. It lived sometime in the 1970s just south of Lizard Island on the Great Barrier Reef. It swam in the company of two others: gold-striped butterflyfish whose shiny bodies looked like someone had placed a dime on a pat of butter. The smallest butterflyfish, my butterflyfish, was half the size of the others but led the trio as they foraged on the reef. When other fish strayed close, my butterflyfish tilted its head toward the sand and prickled its spines in aggression, a guardian. It was the most aggressive of the three toward others, the most on edge toward strangers.

The three fish swam like this for two hours, which we know because a marine biologist followed the trio and wrote down what he saw. After two hours, the marine biologist photographed my butterflyfish and then shot it with a .303 powerhead, an explosive device that concusses the fish to collect it as a specimen. This was, and often still is, common in conservation biology. Scientists could not fully study a fish through a photograph or a tiny sample of its fin, so they had to take the whole fish, dunked into ethanol to ensure it would not rot. It was taken because it looked different, not like a known species but a blend between two different ones, like a hybrid.

I learned about my butterflyfish in a scientific paper from 1977. The scientists described five hybrid butterflyfish, each the offspring of a different combination of species. Each hybrid's headshot was accompanied by photographs of its putative parents and a chart comparing its measurements: length of head, depth of body, number of spines, and so on. I read about the first four hybrids impassively, flipping through photos of the fish to see how the spots and stripes of their parent species shrank, blurred, or were lost altogether. All described the hybrids after death, except for my butterflyfish, whose description included a brief observation of how the fish swam that day on the reef, guarding its companions. It was the only instance in the study that seemed concerned with the actual living fish, not just its appearance or parentage or hypothetical fertility. The description caught me off guard, and I found myself wanting to know more about this small, headstrong butterflyfish. I wanted to know why it was so much smaller than the others. I wanted to know how it came to keep the company of two gold-striped butterflyfish, if they were related or simply happened upon one another on the reef. I wanted to know more about how it had lived its life.

It feels risky, even objectionable, to identify with a hybrid fish, considering how a century ago I might have been considered a hybrid too, how recently Western science attempted to split human races into separate species, how miscegenation laws were only ruled unconstitutional in 1967, how many people on bleak corners of the internet might still leer at my birth. But sometimes what is uncomfortable can also be what feels most familiar, and the closest to home. The first time I saw someone describe mixed-race people as "hybrids" and "half-breeds" was in middle school on a Neopets chat board. The racism seemed so removed from the life I led in ostensibly liberal suburbs, and almost comical against the yolk-yellow backdrop of a virtual pets website made for children. I joked about it for weeks with my friend Saya, who is also half-Asian. But this is all to say that I was probably twelve when I first thought about myself as a hybrid, and perhaps the association never left.

The line between scientific jargon and slur has always been slippery. "Hybrid" came into use around the year 1600, intended to describe the offspring of plants and animals of different species. But eventually its meaning would spill over (offensively) to mixed-race people and (neutrally) to mixed-fuel cars.

The eugenics movement, which emerged in the early twentieth century when a British natural scientist read Charles Darwin's theory of natural selection and decided the principles of selective breeding could be applied to humans, laid part of the groundwork for anti-miscegenation laws. Just decades earlier in the US, sex between white and Asian people was deemed a threat to public health. Chinese women in particular were prohibited from entering the country on the false grounds that all were sex workers and carried venereal diseases that could spread to white communities. The sex was a concern, but so were the hybrid offspring, which certain scientists believed could constitute a new, separate racial species that would eventually render itself as sterile as a mule.

Of course, mixed-race people are not actually hybrids in the sense that the butterflyfish off Lizard Island was a hybrid. I don't want to personally reclaim the word "hybrid," yet I can't help but see the parallels between us. I read news stories about the discoveries of hybrids with charming names: strange narlugas, pizzly bears, sturddlefish. Yet these stories carry a warning. As we humans shape the world and scramble ecosystems and melt the Arctic and intermix animals that might otherwise never have met, hybrids are on the rise. I think of the hopes placed on my body before my birth, and the way I have fulfilled or failed them. My mom used to tell me and my sibling, Sophia, that one of the reasons she married our dad was because he is tall, and she wanted us to be tall, unlike her. The experiment failed: the tallest of us hovers around five foot three. My dad used to tell Sophia and me that his mother, our grandmother, considered half-Asian babies to be the most beautiful babies. At least this experiment succeeded: I was, I think, a very beautiful baby.

In my childhood, my family visited O'ahu almost every summer. My dad was often pulled away for work, so my mom would take us to the beach, where she would watch me and Sophia build sandcastles and leave me to the reef. I swam with a laminated dichotomous key of "Hawaiian Reef Creatures." It was glossy blue and had a fantastic grid of every potential fish I might see, each labeled with its two names: Hawaiian and common. The butterflyfish were easy to spot, clustered together in the top left corner in a riot of yellows and whites.

In Hawai'i, we ate dinner in the condo most nights, microwaving Lean Cuisines and watching cable. But on weekends we dined out at hotel restaurants, where the servers—who often looked much more like Sophia and me than the mostly white families seated throughout the restaurant—would ask my mom about our father. "He's white," my mom would say, and the servers would smile and tell us we were *hapa,* a Native Hawaiian word that I understood to mean part-Asian. I began to feel at home in *hapa,* comforted by the cohesion a single word can offer and relieved to no longer be a half- half-.

In college, I learn that *hapa* is the Hawaiian word for

"part," as in *hapa haole,* meaning a person who is part white and part Native Hawaiian, meaning people like me had stolen the word for ourselves, meaning I was not *hapa* after all.

The father of taxonomy—why must fields of science have fathers?—is Carl Linnaeus, who named more than twelve thousand species. He devised a binomial naming system, where each species on Earth would have a name in two parts: first, the genus, and second, the species. Many of these species already had names, of course. The Native Hawaiians knew butterflyfish—they called them *kīkākapu* and *lauhau*—before Linnaeus named the genus *Chaetodon.* In Linnaeus's system, organisms were named as they were "discovered," meaning, largely, by white men.

When scientists described the hybrid butterflyfish, they did not give it a name. Hybrids do not have Linnaean names because they are not distinct species. Many hybrids are unable to bear fertile young, if they can reproduce at all. They are expected to die out. In Linnaean taxonomy, hybrids look like algebra—two species names adjoined with an ×. A hybrid between a threadfin butterflyfish and a lined butterflyfish becomes *C. auriga* × *C. lineolatus.* These names define hybrids by their parentage, not their individual existences. There are charismatic exceptions, of course—mules and ligers, narlugas and pizzly bears. But butterflyfish hybrids are elusive and serendipitous, seemingly unlikely to become a species of their own or to replace the species they descend from, and so we do not give them permanent names. If they exist in perpetuity, it is in deference to their parent species, the only new space carved out for them marked briefly by the ×.

This × unites many of us hybrids regardless of our mixes. We have stood miserably through attempted genealogical dissections by a guy at a bar or a guy at work or a guy, any guy, really. We have seen the same blurred composite photo of a beige-passing woman that we were told was the mixed-race future of America and wondered why she looked so white. We have felt like a stranger in what we are told is a homeland. This × is intangible and, on its own, technically meaningless. But it is not meaningless to me. It is the only thing we know is wholly ours. We will never be caught between worlds, as long as we have ×; it is our world. When I read *C. auriga* × *C. lineolatus,* the first thing I see is the ×.

"What are you?" is an act of taxonomy, even if the asker does not realize it. It is the question the scientists asked of my hybrid butterflyfish. The question my SAT forms asked of me before I opened my test booklet to write an essay about whether people should accept unfairness as a condition of becoming an adult. The question strangers asked me in the malls of my childhood, peering over my bowl cut to see if a legible pair of parents might suddenly appear. I have lived my life dogged by The Question.

I am lucky that The Question is the most common vector of racism I encounter. It establishes me as something inscrutable—a strange amoeba in a petri dish, never seen before in this pond. In any other context, I would appreciate the reminder that I am an organism like any other, we people and pigeons and bacteria experiencing homeostasis on the sidewalk. But unlike real science, the driving question here is not in pursuit of knowledge but objectification. The Question does not understand me as a person but as an object—not *who* I am but *what*.

As the years pass, I have become convinced that the people asking me The Question are looking not for an answer but for confirmation. I know this because when I do tell these people "what" I am, some of them argue with me. "You're not Korean?" a Lyft driver asked me once, incredulous. "I could have sworn you were Korean. You sure you're not Korean?" In 2019 in New York, everyone thinks I am Korean, so much so that I begin to ask my friends if they're experiencing it too. When my friend Angela, who is Chinese, says she has also been recently, abundantly mistaken for Korean, we hypothesize that Koreans must be "in," at least in Brooklyn. My friend Hannah, who is Korean, says this is around the time more people started asking her if she is Korean; previously, they always guessed Chinese. I develop a casual taxonomy of my own: what the race people mistake us for reveals what they want from us. If they say we are Korean, it means they find us beautiful.

If they say we are Chinese, it means they want us to go back to where we came from. They never ask if we are Japanese, unless it is in a list given to us in some kind of tasting menu of East Asian ethnicities.

It is tempting to write against The Question, and advocate for a future in which mixed-race people are no longer intriguing ciphers to be unscrambled on the sidewalk, in which we can simply exist, unbothered. It would be even simpler to dismiss these questioners as random asshats who need to get a life and stop interfering with mine. But I can't. Because whenever I meet a mixed person who looks something like me, I want to ask them The Question. I want to know what kind of Asian they are. I want to know how their parents met. I want to know what words they use to identify themselves. I want to know how close or distanced they feel to their own whiteness. I want to ask them the questions I don't want strangers to ask me. In other words, I am also the asshat. I can never abandon The Question because I am endlessly curious about our shared hybridity. Maybe it's because I grew up longing for role models, surrounded by dozens of white and Asian families like mine but no mixed-Asian adults. We were all children, a new and blurry generation, all of us ogling the oldest of us for some glimpse into our uncertain futures, some idea of who we might grow up to be.

For some time, whenever people asked me The Question, I ignored them, my pace quickening in a huff. I pretended not to hear. Maybe now I would answer, my steps slowing until I am planted on the sidewalk, my stance so wide that people must walk around me. "What are you?" they ask me. I look them in the eye and tell them: "x."

When I first introduced my hybrid butterflyfish, I said it swam in the waters off Lizard Island, near Australia. But I don't want to say Lizard Island anymore, because that name was given to the island after Captain James Cook visited once, saw what he deemed a remarkable number of lizards, named it, and left. Before Cook arrived uninvited, the island was named Jiigurru by the Dingaal Aboriginal people who have lived on the island for tens of thousands of years. From its perch on the Great Barrier Reef, Jiigurru has since become an internationally regarded research station for scientists studying coral reefs and the creatures who inhabit them.

One of the men who described the hybrid butterflyfish at Jiigurru was John Randall, a white scientist who spent most of his career in Hawai'i and other tropical islands. Randall, who worked at a time when many species of reef fish still lacked scientific names, described more fish species than any other person. He was one of the first scientists to dive with scuba gear, which opened up new realms of the ocean and allowed him to collect the fish obscured in its depths. They called him Dr. Fish. Randall named 834 species of fish, most of which live by coral reefs near tropical islands he is not from.

I first learned of Randall shortly after he died, in the spring of 2020. He was ninety-five. One of his obituaries mentioned his wife and longtime collaborator, Helen Au, who survives him, and two of their children. In one old profile of Randall, I see a black-and-white picture of the young couple, Randall and Au, looking at a poster of parasites found in a surgeonfish. The image disorients me the first time I see it: the tall white man in a Hawaiian shirt with gently parted brown hair and glasses and the shorter ethnically Chinese woman with permed black hair. They look just like my parents. I don't mean this in the sense that Randall resembles my dad or Au resembles my mom but rather that they fit the same, eerily familiar mold. I have to blink to shake the resemblance.

I know this is a mold because I have also been perceived as a part of it. When I was younger and looked more like a girl, people would sometimes mistake my dad and me for a couple. We knew this both indirectly, how eyes would shift between us, reflecting back unease or judgment, and directly, such as the time a hotel clerk tried to sell us a romantic sunset cruise. We developed a knee-jerk dismissal of these charges—both of us laughing—and I told myself that this was just another classic half-Asian experience, and eventually I could write a funny essay about it. When I posted about one of these incidents on Facebook, my half-Asian friend Anna commented, THE WIFE THING ALSO HAPPENS TO ME AND MY DAD!!!!!!!! it feels weird and bad. I always wondered who to blame for these misplaced, unwanted assumptions: the people who made them or the unmissable abundance of older white men around the world who were dating younger Asian women.

Growing up in the Bay Area — a mixed-Asian mecca — I couldn't help but compare my friends' parents to mine. At one of my friends' houses, we ate with her white father at the table as her Asian mother cooked in the background, only joining for the last third of the meal after all the dishes had been cleaned. At another friend's house, her Asian mother made us steaks and teased her white husband for his inability to handle spice. Another friend's white dad was almost twenty years older than her Asian mom. One time, at some school function, I watched as our parents all chatted together, all our white dads looking infinitely older than our ageless Asian moms, and us mixed-Asian pre-teens watching them. This kind of racial tabulation became a compulsion for me. Maybe on some level I believed that if I studied enough of these couples, one day I might feel assured that my parents' marriage was the good kind, whatever that means. And this is the part where I might tell you how my parents met—but I don't want to, because I have to keep some parts of my family to myself.

According to his many obituaries, John Randall was universally beloved. He was a groundbreaking marine biologist, a meticulous taxonomist, a generous teacher, a wonderful friend. No one has a bad thing to say about him. And then I find a thirty-page memoir Randall wrote in 2001, in the later stages of his career, and because I am shameless and nosy and want to learn more about his relationship with his wife, I scroll through. I stop when I see figure 2: a photo of Helen, above a caption identifying her as "of Chinese descent." This annoys me—obviously none of the photos of the white scientists are captioned "white," and did Randall himself write the captions? I read more closely only to come across a passage in which Randall recounts admiring the naked chest of a fifteen-year-old girl from what is now the nation of Kiribati, a girl Randall describes but does not name. She was assisting Randall and the other scientists with their research, collecting butterflies, and while she was collecting butterflies they were staring at her body. I reread this paragraph twice, first confused and then repulsed, trying to understand what had possessed the seventy-something-year-old Randall to include this memory in a quasi-memoir of his ichthyology career published in a scientific journal. And even though I know about the many fish he named, the field of science he shaped, the people who loved and learned from him, this is now the first thing I think about when I think about John Randall.

My friend Will, who has a Japanese dad and a white mom, works in conservation biology. He studies the gut microbiome of killer whales in the Salish Sea. I often envy Will, whose job includes spending long and beautiful days on a boat studying animals I also love. Sometimes I wonder if I would be so preoccupied with all this if my parents' races had been reversed, my dad Asian and my mom white, the less-charged pairing. Would I not have fixated on this for years? Maybe, no longer racked with all my personal paranoia, I would have channeled my anxiety into something productive. Maybe I would be out there with Will, saving whales.

Sometime during the pandemic, my partner, T, and I are eating beef noodle soup and talking about a new book we have just read about three women: two white, one half-Asian, half-white. I loved the book, but I am complaining about the moment when the Asian woman's parentage is explained by one white person to another—Chinese mom and Jewish dad—like a caption, a specimen ID.

"Why can't she just exist without explanation?" I complain, and as I complain I know that I am being a hypocrite; if her parentage wasn't given, I would wonder what her mix was, whether it was like mine and T's (we are the same mix, and often joke we are in a mono-biracial relationship). I tell T that I am complaining only because I am writing this essay and thinking hard about how other people see people like us. I complain about how often the world feels like it is owed an explanation of how we came into it. T asks me, almost as a prompt, how I might write about these hybrid butterflyfish the way I wish someone would write about me, with no interrogation or speculation of its parentage or guesswork on how it came to be. I could end on that anecdote, but it doesn't feel satisfying enough.

Maybe I could end with a text conversation with my friend Arya —

lol can a white person be Asian tho
I guess we are
Love how biraciality queers the paradigm

we're really breaking down barriers/walls/ceilings
paving the way for more white people to be Asian

Or I could bring it all home with a scene from the time T and I made our landlords and their children — half-Asian, half-white — dinner, and how secretly thrilled I was that this dinner meant, at least subliminally, that T's and my relationship, and our existences, could offer these children a window into their future, what it means to be mixed and grown. But half an hour into my conversation with my landlords' teen daughter, I realize she couldn't care less about our shared mixed race. Maybe she doesn't think about hers at all; maybe it's no longer a big deal.

Maybe I could blend all these scenes into a montage of T and me coexisting with our mixed community, marveling at how good it feels to carve mixed-race pumpkins, sauté mixed-race vegetables, text mixed-race texts, grill mixed-race pizza, shoot the mixed-race shit. Maybe these moments teach me that this joy does not come from being around people who look like you but from people who are irritated in the same ways. Maybe home is the people who hear your rants and nod, because they know. Maybe complaining to someone who gets it is one of the purest comforts on Earth. Maybe it is less about our shared backgrounds than it is about our shared irritations, obsessions, grievances, fears, resentments. We are still dissecting ourselves and how we came to be, but now we are the ones asking the questions.

We Swarm

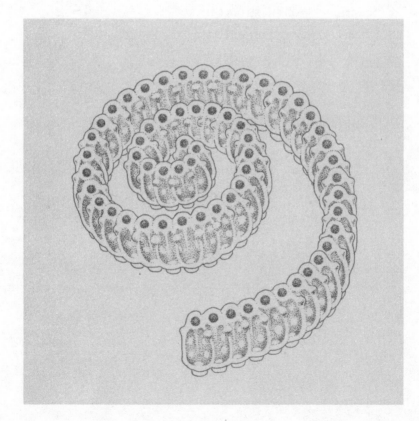

Two Aprils ago, a humpback whale stranded on Jacob Riis Beach in New York and died belly-up: white fins splayed in the sand like a snow angel, grooved throat mounded as if still holding a gulp of air. Its death made the local papers. The 28-foot-long whale was the year's first reported stranding in the area. Its body appeared pristine, at least as far as dead whales go. A week later, the whale was buried, and its beachfront burial also made the local papers.

When I read the whale's brief obituary, I squinted at the photo, trying to figure out where on Jacob Riis it had stranded. I go to Riis Beach often, and I wanted to place the whale somewhere on that landscape. I wanted to know if it had died somewhere I had walked, or, in the future, I wanted to walk across the stretch of sand where it had died. But the picture was placeless, the whale a blip on a stretch of empty sand whose only landmarks were a few distant dunes and a weathered skyline of wood pilings.

I have seen many creatures stranded on Riis Beach that were not large or notable enough to garner press coverage. I remember the holographic sheen of a still-inflated man-of-war, shining like a bubble on the sand. The hollow, chestnut husk of a horseshoe crab, a chunk missing from its side and a pristinely reticulated tail that shuddered when I picked it up. A motionless mole crab with a belly full of tiny, sherbet-colored eggs. These animals died as the whale did—carried too close to shore by warm waters or riotous currents—but their strandings would only make the news if they occurred en masse: a parade of men-of-war in the surf, hordes of overturned horseshoe crabs, a mountain of molted mole crab shells gently lapped by waves. In the headlines, these creatures never "stranded"; rather, they "washed up." Why was that? Perhaps stranding suggested the creature was worth noticing, was worth saving.

The National Oceanic and Atmospheric Administration, or NOAA—the governing body of matters like these—has a dedicated network that observes the strandings of marine mammals and sea turtles. There is a bureaucratic system for people to report strandings, people and agencies to call as if they were next of kin. There is even an app for whales and dolphins. There is no stranding network for smaller creatures without faces, like mole crabs or men-of-war, whose bodies erase themselves by washing away or dissolving back into sand.

Several years ago in September, I saw shimmering heaps of gelatinous blobs strand on the shores of Riis Beach. Everyone who went near the water that day noticed the blobs, including my friends, my crush, my crush's crush (who, sadly, was not me). But they were reluctant to inspect the blobs, which had amassed in nubbly, sparkling dunes. "Is it alive?" someone asked. "I don't want to get stung," another replied, and they walked away from the water to spread their towels and single, battered umbrella far away from the lapping waves.

"They don't sting!" someone in a flame-licked Speedo shouted at us from the breaking point of the waves, just before diving in. I followed their voice, the slimy pebbles slapping my stomach with each new wave before I plunged my whole body into the water. The ocean felt strangely viscous, almost sticky, and each time I surfaced I saw glassy orbs the size of boba drip off my arms. I could feel them as I treaded water, my hands parting the creatures like a beaded curtain, only for the blobs to reform in their watery swarm. I paddled near the person in the Speedo, whose friends were speculating on the possible identity of the goo balls.

"I bet it's baby jellyfish," said someone in a green bucket hat.

"But they don't sting," Flaming Speedo countered, intent on defending the nameless creatures.

"I think it's fish eggs," said another, the ocean obscuring any swimwear they might have had on and occasionally splashing into an open cup of rosé they held, precariously, above the surf. We continued spitballing together, pausing only to jump with or dive under the largest waves or shout reassurance to would-be swimmers on the shore, who prodded the blobs with sticks and tentative toes.

"They don't sting!" we chorused, having no formal expertise aside from our shared sensation, our bodies touching blobs, touching bodies, none of us hurt. We twirled in the slick, batting the gummy blobs back and forth like children in a pool. They were so numerous they seemed even to dilute the crashing force of the waves, leveling the peaks of the afternoon swell into something almost languid. It was strange to swim this way, held afloat by a cloud of creatures so densely clustered it felt impossible to sink.

I touched down in the sand and picked up a handful of the orbs to examine them. They didn't look like jellyfish, at least not in the conventional sense of a bell-headed medusa with trailing tentacles. They seemed too big to be fish eggs, too symmetrical to be tatters of men-of-war or any other larger thing I'd seen wash up before on Riis. They looked like raindrops, or tears, water in a state of falling. I couldn't tell if they were dead or alive. I held one up to the sky and its dimpled gelatin muddled light like a prism, turning sunbeams into deliriously electric blues, cherry-blossom pinks, kelpy greens. I threw handfuls of the blobs in the air above me and the droplets filled the sky, shredding sunlight into rainbows.

Days after I left the beach, the blobs had taken over my mind. I Googled hopelessly: "small translucent circles stranded on beach" and "clear gelatinous lumps on beach no sting" and "what is this goop I found on the beach" and "what do baby jellyfish look like" and "weird stuff washed up Rockaway?!!" But I found no record of the blobs' presence outside my memory. No one had written about them or even guessed at what they might have been or where they came from. Perhaps this was the fault of the blobs (too benign and formless to make a good local news story) or the fault of where they had stranded (a beach infamous for its queer visitors, not wildlife sightings).

I thought about the blobs every so often for several years, until I decided to email a park ranger named Dave — supporting my ongoing theory that all park rangers are named Dave — to ask if he knew what the blobs I saw years ago might have been. I had no concrete evidence of the blobs, no better description for them than blobs. I almost felt stupid for asking — who would remember something so inconsequential and amorphous from so long ago?

Ranger Dave asked for a photo, but I hadn't taken one. I told him they were firm yet gelatinous, ovoid and transparent, and he said many things that lived in the ocean looked like clear orbs of goo. I asked Ranger Dave if the blobs might have been a kind of salp, a colonial animal that spends part of its life surrounded by clones of itself. He said it was a good guess, but without a photo, the blobs' true identity may remain forever unknowable to us. But he would be happy, he added, to speak to me about the 28-foot whale that had washed up in April—had I seen it in the papers? Perhaps I would never learn what the blobs were, or even what they could have been, but I wanted to have a name for them, because they mattered to me. So I gave myself permission to remember them as salps.

Despite their dour name (no one ever looked at something beautiful and named it salp), salps are fantastical creatures. If you dive deep enough, some salps even glow. On shore, they look like beads of clear Jell-O. But in water, they exist in pulsating chains that can curve like a snake or coil like a snail's shell. These chains are made up of hundreds of identical salps joined hip to jiggly hip. Each clone is a distinct, barrel-shaped individual, yet all together the colony of clones makes up a single salp, attached and moving as one. Many chains can grow as long as 20 feet, drifting through the ocean like giant quartzite bracelets. This is to say that individual identity is confusing for a salp, creatures for whom the notion of selfhood exists in the plural. For a salp, home is the rest of its salp.

Although salps have no limbs, no discernable muscles,

and no clear agenda, they still manage to move their bodies from one part of the ocean to another. Salps move by jet propulsion, each individual sucking in water at one end of their body tube and clenching that body to shoot the water out the other end. A salp chain does not move in one great coordinated effort, with each individual salp synchronizing its jets. It may seem surprising; surely a winding chain of individuals would move more efficiently if they sucked and clenched and spurted in perfect timing. After all, this is how a jellyfish moves, in rhythmic bursts of speed and stillness, waiting for the rest of its body to catch up. But salps allow each individual to jet at its own pace in the same general direction. It is not as fast as coordinated strokes, but it's more sustainable long-term, each individual sucking and spurting as it pleases. Slow and steady, say the salps. It doesn't matter how fast we go, only that we all get there in the end. We may all move at different paces, but we will only reach the horizon together.

Much of the time, salps keep to themselves, out of sight. They lurk in the deep where the water is cold and rich in nutrients, thousands of feet below the surface. There are no boats here, no nets or wrecks, so the salps can jet around in peace, gulping in particles of food that get caught in their mucous net. Some species of salp migrate up to the surface at night to feed on phytoplankton, to encounter other salps and multiply and breed. When the sun inches into the sky, the salps return to the deep before the glassy walls of their body shimmer in the rays and attract a predator.

Sometimes, when the currents generated by the wind and rotation of Earth lift deep, cold water to the surface of the ocean, salps rise in swarms. These cold plumes act like fertilizer and can feed enormous populations of phyto-plankton, which feed hordes of salps breeding so rapidly that they form blooms, billowing salp clouds. These blooms are ephemeral and dizzying. They rise and swarm in the billions, cloning themselves until they cover huge swaths of ocean in a glossy haze. In 1975, one swarm of thumb-sized salps covered 38,600 square miles of waters off New England. The salps lingered for months, eating the micro-scopic plants that drifted in the water and excreting fluffy squares of poop that sank, quickly, away from the light.

Every June in New York, we swarm. We come from all around, on trains from other boroughs and cars from upstate and bikes over bridges that seem to quake, throt-tled every few minutes by subway cars careening into open air. However we come, we always recognize one another, limbs stuffed in mesh and netting and leather, teeth bared, nipples out. Our shirts, if we wear them, are emblazoned with the conditions of a world we would rather live in: without TERFs, without ICE, without imperialism. We hold cardboard signs that often prioritize the specificity of our message over easy legibility, and we crane our necks to read the message in full, because we know it will be worth it: SEX WORK IS WORK airbrushed onto a crinkly pink dress and EATING ASS IS THE ONLY ETHICAL FORM OF CONSUMPTION UNDER CAPITALISM in bubbly letters on black cardstock. We meet in a part of Manhattan many of us have no business in, a patch of green surrounded by glass-fronted stores and metallic offices, and once there, we grow larger, friends finding friends and water-getters winding their way through an obstacle course of bodies. We swarm because we are full of the joy of being together, full of anger at the systems that exclude or endanger us, full of hope for the possibilities of the future. There are no metal barricades to keep us from unspooling onto sidewalks, brushing up against luxury soap stores that have steeled themselves against our arrival.

The next day, the city will erect barricades and garlands of cops to oversee the capital-*P* parade, the one first named after liberation and then later rebranded as Pride, the one sponsored by the city and Bank of America and Amazon and other institutions that do not care about us unless we become something to monetize. The next day, cops will roll in on side streets in terrible white cars with rainbow-striped decals of the NYPD logo to escort a behemoth of a float from Gilead, a pharmaceutical company that manufactures Truvada, a daily pill to prevent HIV, and charges $1,500 to $2,000 for a monthly dose that costs $6 to make.

But it is not Sunday yet. It is Saturday, the day of the

Dyke March, which any of us will remind you is a protest, not a parade. The Dyke March has no official permit, no corporate sponsors, and no invited police presence. There are no steel barricades dividing the marchers at intersections, just volunteers interlocking hands across one avenue in Manhattan to allow our ecstatic masses to march down the gray grid of Midtown in one great, unbroken chain. At some point—we are usually on time, though always welcoming of latecomers—we spill into the street and start moving downtown. We pound pavement, albeit softly, walking at the pace of our slowest companions, stopping when we need to tie a shoelace or say hello to an ex or ask a friend if that person in the leather harness—no, the other person in the leather harness—is the same person we kissed by the coat check at the lesbian party at the bar decorated like a subterranean cave, where stalactites dangled phallically from the ceiling.

One year, we swarm into a Panera Bread, me and a handful of people I have just met, jetting without hesitation toward the back of the store. We erupt in the bathroom, shedding pants before we've closed stall doors, and when we hear a manager shouting that we must leave, we offer to buy water and muffins. When he says no, that the bathroom is closed to everyone, we shout that this is actually not some sacrosanct hall but a Panera Bread, that we have no other way to pee, and we barricade ourselves in the bathroom until every last one of us has peed, has fixed our hair or makeup or washed smeared glitter off our cheeks, and, with one great breath, we open the door, and for a second I meet the eyes of the person working the register, their face blank, and I know we have made their day more difficult so I mouth, "We're sorry!" as we scramble out of the Panera and back into the street.

Unlike Pride, there is no strict demarcation between marchers and spectators; we move fluidly on and off the streets. The few permanent spectators are often men who've come to protest, holding banners saying something about Christ and sin and the promise of our eternal damnation, but there is always a group of us surrounding the man like a bubble, holding signs of our own: THIS GUY NEEDS A HOBBY! Half the joy comes from watching our own, turning around to see everyone who walks behind us. Many look like someone I might know or see at a party—youngish dykes in bowling shirts—but my eyes always drift to those I might not have encountered were it not for this day: silver-haired couples in matching blue Hawaiian shirts, steel-spiked butches on motorcycles, dykes carrying babies in slings and in strollers. One year, I see an older woman, short white hair, holding a sign. IT IS OK TO LIVE A LIFE OTHERS DON'T UNDERSTAND. She drifts at her own pace, slow and steady, as younger marchers weave around her. I feel overcome by the urge to go up and thank her. I want to march in service to her. I think about what she must have sacrificed to make it this far. But my group of friends are marching quickly, to keep up with the drummers, and so I turn around and run ahead.

Though our swarm looks different every year, familiar faces in new outfits and new relationships and situationships, old marchers moved to new cities and new marchers fresh out of NYU, we always end the same way, the crescendo of our shouts passing under the triumphal stone arch in Washington Square Park and our bodies streaming across the cobblestones of the courtyard as we watch some of us—the brave, the hams, the unusually germ resistant—take off our shirts and jump into the waters of the fountain, so ecstatic we have to cool down. There, in the water, we splash one another, kiss one another, hold one another, all our soft parts jiggling as we pulse together in one final swarm before trickling off to go our separate ways.

Salps do not spend their entire lives in gigantic chains or spiraling colonies, surrounded by clones of themselves. The creatures alternate between colonial and solitary stages that look entirely different. A solitary salp, unlike the colony I encountered at Riis, resembles a hollow tube, made visible only by the golden peppercorn of its gut. Solitary salps grow their future colonial selves inside their bodies, creating a chain of genetically identical clones strung together like pearls on a necklace. As the single salp grows, so does its internal chain of clones, until the chain is long enough and big enough to break away from the original body.

Once liberated, the chain of clones tumbles into the ocean, where they swim as one, bending and wavering like a liquefied spine. Within the chain, each clone has a single egg that, when fertilized by gametes from a nearby salp, the clone will carry until its young embryo is old enough to swim away and start life as a singleton. Once its embryo is gone, the clone will grow testes that spurt out plumes of sperm, which scatter in the water to fertilize the eggs of other clones. This is how salps swarm, one chain producing hundreds of new chains, how they overtake vast patches of ocean in a way that upends the ecosystem. When an ocean teems with chains and helixes and whorls of salps, it is said to be blooming.

Always alternating between life stages that barely resemble one another, salps long eluded scientists. Salps were first described in 1756, by men who did not understand how they lived or how they reproduced. Not that they—or many scientists who came after them—tried very hard to understand. With no head, no brain, and a body that slips easily out of a hand, salps were hard to track down and harder to study, appearing in enormous, unpredictable clouds and disintegrating quickly. Many zooplankton ecologists try to avoid salps in their sampling because of their messy abundance, their mysterious and complicated taxonomy, the way their glassy bodies shatter in nets. When salps moved alone, they went unnoticed. When they moved in swarms, they posed no great threat, it seemed, only inconvenience. For centuries, only one thing seemed clear: the salps, wherever they went, were unwanted.

So there is almost no long-term historical record of the salps in any ocean. Few data sets go back more than twenty years, and the ones that do record salps only in these ephemeral explosions of abundance. Scientists only took notice of them when they gathered in such great masses that they made themselves impossible to ignore. Perhaps you would not see salps if they did not form blooms. Many scientists consider salps a nuisance species, because, in swarms, they can take down a fishing net or stop a ship. In 2012, a swarm of salps clogged the water intake system of the Diablo Canyon nuclear power plant, the last in California. It is amazing what salps can do in community.

In recent years, some scientists have deemed salps and other forms of gelatinous zooplankton—creatures whose body is more than 95 percent water—an unwanted sentinel of ecological disturbance, a future ravaged by climate change. Sightings abound of this surge in swarms and the trouble they cause: deluging fishing nets and invading the cooling pipes of the nuclear-powered aircraft carrier USS *Ronald Reagan*. It is true that scientists have recorded an enormous number of blooms of salps and jellies in the past two centuries. But a group of research scientists argue that this "perceived increase in the number of jellyfish blooms may be a case of shifting baselines," as they write in a paper published in the journal *BioScience*. "The public perception is shaped in the absence of a historical baseline and through a lack of continuity in the collective memory."

The day after the Dyke March, meaning the day of Pride, when swaths of Manhattan become untraversable, I go to Riis. The beach was originally named after Jacob Riis, a muckraker who photographed New York immigrants in cramped and squalid tenement houses, exposing the cruel living conditions at the turn of the twentieth century. Riis died in 1914, the same year this beach became his namesake. He never lived to see the far eastern end of his beach become a haven for queer and trans people, which is perhaps for the best. Riis did worlds of good for the poor and destitute, but he did describe "Orientals" as sinister and Italians as unsanitary, so one can only imagine his stance on homosexuals.

We swarm on the gay stretch of Riis, umbrellas standing neck-to-neck, towels overlapping, microclimates of speakers blasting Mariah and techno in a thudding, electronic din. The beach is not squalid, but it is far from pristine. The tides seem to funnel all the trash from the mile-long beach to the gay end of Riis, where plastic bags bob like crinkled jellyfish. Beyond the sand, the trash cans overflow, releasing the metallic wings of torn chip bags and their confetti crumbs into the dunes, a banquet for seagulls. And behind it all, pressed up against the throngs of queers and our wafting flotsam, sits the abandoned sanatorium for children with tuberculosis, enrobed in chain-link fencing and garlanded in barbed wire. It seems hardly coincidental that the least scenic stretch of Riis Beach came to belong to queers. The straight people apparently did not want to spread their towels near the rockiest end of the strip, did not want to sun under the drab brick of the sanatorium, and so now it is ours.

Riis on Pride is the opposite of a parade sponsored by a pharmaceutical company. Riis is where I can see everyone I love, or at least everyone I love who is queer and lives in New York, which is a great portion of the people I love. On Sunday, at Riis, I see Olu, who tells me about her crush—she has a new one every month—and CV, who most likely drove my partner, T, and me there, who always manages to get sunscreen in at least one eye, along with Caroline and Indigo, if Indigo wakes up in time, and in the distance I see my ex, whom I sit with as T talks to their ex, and inevitably Trace wanders by, always cruising, platonically speaking, crisscrossing the sand in search of scattered friends, and I see Kiyana and Rachel in their silky ivory tent by the hamburger stand, and I see Shirley and Lila, the cutest best friends, and I see Joey and Mads and Mer piled up under a red umbrella and Lisa starfished in her recliner, and I see my old barber Alana, her salt-slicked hair always perfect, and I see the ice-cream scooper I matched with on Tinder sitting with someone I met just yesterday on the street at the Dyke March and then again on the subway home, our meetings like bookends of a perfect day, and I see the couple I met at the march the year before when they were first falling in love and I was falling out of it, and I see Marion flinging herself into the waves, hours after the water stole her sunglasses, and I see Riis's regulars, the older person in the black thong and the inflatable bull and the person who always sits fenced in by Barbies in drag and my favorite DJ and my favorite DJ's favorite DJ and the lesbian potter who lives in my neighborhood and the group of friends I met at a launch for a novel about a lesbian taxidermist and I can't remember all of their names but I know their faces and I wave and they wave and then, inside the swarm, I spy an empty patch of sand, and I run to it. I roll out my towel. I take off my shirt, my body hair crunching with salt, and I kiss T, who insists we put on sunscreen before we break out the paddleball, and I close my eyes as they coat my back in cool white slime, and I feel safe. It feels like pride, sure, but it also feels like liberation.

At Riis, we pass the speaker. We lend an umbrella. We share a six-pack, rosé, SPF 30, SPF 70, aloe vera for the pale or carefree. Goggles, a hat, a shirt when the clouds slip over the sun. A beer, the rest of a sandwich. Paddleball, a deck of cards, a phone number. We put our lips on the mouth of the inner tube and we blow. We swim out to the sandbar, hold hands, and dig our feet in so when a wave comes we are ready. When the sun disappears behind a cloud and the sky grows somber, we unfurl our umbrellas in a murmuring crackle. When the sun emerges again, we cheer. It is sunny again, for now.

For now, of course for now. Soon the sky will darken, the rain may come, and then winter. Some of us may move away, to another borough, to another state. Some of us may realize we are too old to schlep a cooler two hours on a bus to sit among shouting twenty-somethings. Some newcomers will step off the Q35 and see all of rainbow Riis for the first time and fall in love like we did and keep coming back. We are not forever. Riis may outlast us, but it will eventually disappear too, swallowed up by rising seas. But for now we are here. Now is the best day of our lives, until we come back. Now we soak up all this love until it rolls down our backs in salt and we jump, screaming, into the water. Right now we are so in love with one another that we need to cool down, to submerge until everything disappears so that when we surface and open our eyes, we are newly amazed. That it is all still here. That we are all still here.

When I learned that the eastern end of Riis had been a gay haven as early as the '40s, or even the '30s, I was stunned by the longevity of the site, and of what we had inherited. I'd imagined it dating back mere decades, not the better part of a century. The people who first frequented the beach were white gay men, of course. Lesbians (white) came in the 1950s, and in the 1960s queer Black and Latinx people staked a claim to the beach too.

Queer archivists and historians have maintained their own history of the beach, comprised of grainy gray photos with unidentified, smiling beachgoers and fliers for parties that took place in the 1960s. The more official record—histories collected by the government and mainstream media—memorialized Riis in reports of police raids on public sex in bathhouses and vague sociological observations of the beach's queer demographics. In 1974, the *New York Times* announced eleven men had been arrested for engaging in sodomy in Jacob Riis Park. In 1991, the paper described the eastern end of the beach as "filled mostly with homosexuals."

There are also the lovingly taken but anonymous photographs. A group of pasty-to-tanned men standing in a circle, a more pristine brick building behind them. A grainier group shot of women against a washed-out ocean squinting toward the sun. In one photo from the 1960s, a person wearing white heels, a white turban, and a white terry towel draped around their waist clasps their hand to their chest and looks behind them at the camera, brows arched. They are beautiful, like something conjured from the waves.

When I look at these photos, I know without question these people are at Riis. It is not because of the generic expanse of sand or timelessly lapping waves but the familiar hulking brick skeleton that serves as the unintended backdrop of so many of our photos, in various states of decay. The sanatorium appears in the chiaroscuro photos of archives, in the sudden color of 1980s Polaroids, in the corners of our iPhone photos. Sometimes you can see the whole stalwart building, sometimes only its fence, sometimes merely its long shadow. It anchors this place. It has seen us all. We queer beachgoers come and go, but the abandoned children's tuberculosis hospital will always greet us when we return.

Throughout the first half of the twentieth century, children sunned on the decks that wrapped the hospital, taking the so-called heliotropic cure. The city closed the place in 1955, when a microbiologist from New Jersey cured tuberculosis. It reopened in 1961 as the first municipally operated elderly home in the city. In 1985, Mayor Ed Koch announced plans to transfer ten patients diagnosed with HIV/AIDS to an isolated, already vacant wing of the Neponsit Health Care Center. The Neponsit community, affluent and majority white and filled with misunderstandings of how the virus spread, protested, and Koch, who famously failed to take action amid the mushrooming crisis in New York, gave in to their demands. The patients died elsewhere, likely farther from the ocean. The hospital remained open until 1998, when a storm whipped through Riis Beach and peeled away stones and bricks from the building's walls, leaving the sanatorium a dilapidated fixture in the city's queer life.

In recent years, the hospital has been slowly taken over by those who treasure the beach. As I've watched the building crumble, I've also witnessed it become an altar of sorts, for queer lives past and future. In 2018, part of the concrete barricade blossomed into a memorial for Ms. Colombia, a performer who wore dresses festooned with rainbows, dyed her voluminous beard highlighter yellow, and drowned in the waters off Riis one Wednesday morning. A vigil re-created her silhouette in fake flowers and garments pinned to the chain-link, and a muralist painted her name in gem-colored hearts. In the following years, elsewhere on the concrete, DYKES & FAGGOTS RUN NYC appeared in bubbly graffiti. Someone scaled the hospital and painted at the top of one tower QUEER TRANS POWER in marshmallowy all-caps letters, a reminder of the urgency of our softness.

In the ocean, the bodies of salps are often repurposed by tiny amphipods whose glassy claws are speckled copper. One such amphipod, *Phronima sedentaria,* has bulbous eyes with crimson retinas that help it spot diaphanous creatures like salps in the open ocean. Once it spots a salp, *P. sedentaria* uses its grasping appendages to cling to it, hollows out the tunicate with its scythe-shaped front claws, crawls inside the body, and carves out the creature's quaggy insides with its knifelike mouthparts. Finished, it spends its days living inside its new conveniently aerodynamic cocoon, holding its front legs and body inside the salp and kicking its feathery tail legs out the other end. The amphipod lays hundreds of eggs inside the barrel shape of the salp's body, a soft shield that dilutes the ocean current.

This relationship is technically parasitic, but *P. sedentaria* is less a parasite than a parasitoid, a creature that kills its host. The amphipod more or less guts the salp as it takes over its body, but it also allows a strange and wonderful kind of cohabitation, between the living and the dead, or almost-dead. The salp is no longer fully there, but it has transformed beyond a creature to something like a shelter, a home. And, cells still alive, it continues to exist alongside the amphipod and the amphipod's young, both eggs and newly hatched babies too small to leave. It is an eerie but pristine preservation of a body without rot or disintegration. It is, in other words, the closest thing I can imagine to living alongside a ghost.

I have never come close to drowning, but the possibility sometimes comes to mind when I am swimming in the waves. Sometimes at Riis, when a surging afternoon wave comes to us, perhaps more dangerous than we understand it to be, we stay in the water and scream together, alive, hearts pounding, and when the water retreats, we push our heads above water and gape, the cold air inflating us upright.

But that September at Riis, the day I swam with the blobs, their bodies teemed in the water so thick that drowning felt impossible. I knew, logically, this was not true, but each time I kicked, even thrashed, in the water, I touched something alive, or close to it. I felt as buoyant as I'd ever been, held in the water that day. I imagined myself a spineless, fleshless blob, 95 percent water, my body a water body. I imagined ourselves billions of years ago, a time before salps, all of us blobs in some primordial sea, becoming the first microbes. All of us on the cusp of inventing life. The first cells on Earth were more like bubbles, fatty molecules coating foam, and they wavered in and out of existence as they circulated in the ocean. Back then, the whole world was ocean, a globe made entirely of blue seas studded by occasional islands. The only place to live was underwater.

The poet Ross Gay asks if joining together all our sorrows—all our dead relatives and broken relationships, all the moments that make life seem impossible—if joining all these big and little griefs together, if that constitutes joy. As I watched the other beachgoers floating among the apocryphal blobs, all of us strangers until this strange, shared moment, I imagined my body chained to their bodies. My sorrows to their sorrows. My survival to their survival.

As I write this, I have not been to Riis in two years. At least, not the real Riis, the Riis that I know. I did go to the beach this past summer with my tiny pandemic pod and saw other pods in the distance, all of us afraid to draw too close to one another, and it did not feel like Riis. The sun was out, but it felt lonely. I don't know when I will go to Riis again, but I want to, I want it, I want Riis, my swarm of people shimmering together in sweat-slicked murmurations, a sunburnt bait ball on the sand.

I want to see all of my exes—let them come—and to feel a surge of remembrance of the times I loved them most, our moments of greatest connection, the closest we came to imagining a future together, futures like building a home in the South, seeing the killer whales in the Pacific Northwest, growing old in the same town we grew up in. I want to see everyone who has moved upstate or to Los Angeles or grown distant for reasons beyond geography, tangled in their friends and lovers, ex-lovers, lovers-to-be. I want, impossibly and more than anything, to see the people who first found refuge on Riis, who sat in sand that no one else wanted, who lived in spite of the people who wanted them gone or dead. Maybe they imagined what the beach could become, or maybe they had no inkling of the future they were building, how their Riis would continue on and on and on. I want more than exists in the archives. I want to know them, to understand the texture of their lives, who they loved (and who they despised), how they spent their days, what they brought to the beach in bags or bundles, what music they heard or made, how they eked out joy. The tubercular children can come too, and sit among us while they soak in the promise of sunbeams, all of us relearning what it means to be cared for.

I want the breaking of each new wave to collapse the years that separate us into a single timeline, a single day at the beach, where no one is competing for space but where we are together, not just the people who have come to these shores but the creatures too: 28- and 29- and 30-foot-long humpbacks that have come to Riis to die, the last scuttlings of horseshoe crabs, bean-sized clams and the shorebirds that peck at them, the maybe-salps, along with every single strange and gelatinous thing that loses its shape when it leaves the water. I want all our soft bodies pressed together. I want all of us, teeming masses, swirling riot, heaped atop one another until we become something more than summer, more than life.

Morphing Like a Cuttlefish

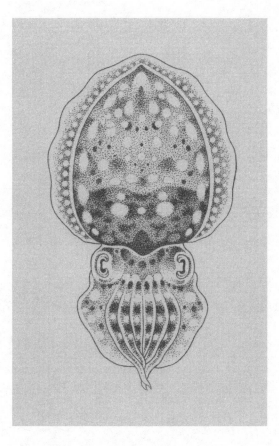

Imagine you are something like a snail. Your shell resembles a gnome hat, pointy on top and fluting on the bottom into a bouquet of tentacles, which is you. Your shell is a part of you, but only technically, an exoskeleton that protects the soft frill of your body. Perhaps you'd rather do without it, but your ocean is dangerous. It is the late Cambrian, about five hundred million years ago, and the seas brim with predators that would love to eat you. It makes sense to ensconce yourself in a shield. You are not a fast swimmer, but you make do, bobbing around the blue and sealing off your body in spiraled chambers that buoy you up and down the water column. Everyone can see your shell, but no one can see the body curled inside it, unless you let them. It makes life less dangerous, though it will always be dangerous, in a way, to live on this Earth.

In the late Devonian, life above you starts to morph, as

slow and certain as a sunset. On land, mites and millipedes trundle around towering ferns and the first seed plants, which will usher in increasingly canopied forests. In the sea, fish are growing bigger and faster and better at eating you. They've learned how to pry you apart from your shell, how to crush it in two, how to drill through it and suck out your meat like a milkshake through a straw. Having a shell seems far less useful, and sometimes a hindrance. Its weight may even drag you down, making you easier prey than something that can flit away in a second or two. So when the species around you shed their shells, you follow suit.

Now you're squishy, naked. You're embarrassingly obvious to anyone who comes near. The bare flesh of your body is tantalizing to others, advertising you as a snack, so quick and easy to chew. You can swim faster, compressing your body into the shape of a torpedo and zipping through the water. Your shell, now sheathed inside your soft body, is smaller and lighter. You can shoot clouds of ink left and right, sometimes even creating shadowy replicas of your own body, but there are still too many close calls.

As you evolve, two hundred million years later, your brain balloons, unlocking extraordinary abilities. Your skin becomes a shimmering screen of pixels, studded with pouches of earth-colored pigments that your muscles pull open and squeeze shut. Clenching this way can lead to a ripple of change across your body. You, cuttlefish, can change your appearance in a fraction of a second. You can stripe yourself like a holographic tiger or freckle into muddy shards of sand. You can make your skin appear to move when you remain still. You can even disappear. Shape-shifting this way helps you evade predators: sharks, seals, fish, and even large worms. But once the danger passes, you may feel tempted to seize this metamorphic power for means beyond escape. When you are not fleeing, what will you become?

As a child, I was good at being a girl. I wore bubble dresses, wedge boots, an unseemly number of berets. My hair grew long, an oily black petal that tumbled into my face. Like many other girls, I waited anxiously for the jumble of features that stared at me in a mirror to rearrange themselves into something beautiful. Puberty came and passed, and I did not become beautiful, or at least I did not feel beautiful, and when people told me I was, it felt like the world had conspired against me to buoy my self-esteem. By college, I'd become a pest, wine-drunk and pulling reluctant friends into bathrooms to explain how the geometry of my face or body was wrong, couldn't they see?

Some days it was my eyebrows (too thin) or my cheeks (too plump) or my stomach (is there such a thing as an attractive stomach?). Some years after college, my baby fat shrank away to expose my cheekbones, and I learned to draw in my eyebrows. I could look in the mirror and see that I made sense as someone, which was such a relief that

I didn't take the time to wonder if that someone was a person I wanted to be.

For several weeks in the first summer of the pandemic I stopped looking at myself in mirrors. It wasn't intentional, but once I realized I hadn't seen myself in a few days I decided to keep up the game for as long as I could. It was frighteningly easy; I had no job, no video calls or coworkers. I washed my hands staring at the bathroom sink and tied my shoes facing the door. I saw myself as hands and stomach, pale legs and faraway feet. It wasn't that I didn't want a body; I wanted a body badly, just one that could change, mutate, evolve. I pressed my chest down until it felt flush with my stomach. I tensed my arm to watch my veins surface like big blue whales, and then melt back into skin. I touched my face so much that a pimple, big and painful, erected itself on my chin. I stopped the game so I could look at it: white-capped marvelous thing.

The rift I feel from my body has made me grateful for the ways I have managed to alter it: tattoos and buzz cuts and piercings. Perhaps if I was not an individual organism but a species, able to prototype myself over millions of years, I might have faith that my body would one day become what it's meant to be. Maybe I would have elongated limbs to skitter atop water or bony flippers to wedge into sand. I might have a layer of gelatin or retractable spines. Maybe the way I understand myself can accommodate the physical changes that only happen on an evolutionary timescale. Maybe my ideal form is one that would be unsuccessful, a mutant too showy or slow for its own good, snuffed out too soon by a mosasaur. But for now, in the brief splendor of my human life, I don't have millions of years to let evolution figure it out for me. I have to start morphing on my own.

Cuttlefish are creatures born to morph. Each one can have up to millions of chromatophores, skin cells that can stretch or squeeze the pouches of pigment they contain. This is how they change color, squinching their mosaic of reds and yellows and browns. Underneath this top layer, cuttlefish have a layer of shimmering organs called iridophores coated with hard plates of chitin, stacked atop one another like felled dominos. When a ray of light strikes a cuttlefish, some of the light reflects back, staggered by each plate of chitin. Iridophores interfere with how certain wavelengths of light are reflected, creating a shimmering rainbow of iridescence. What does not reflect back is absorbed or reflected in the rice-shaped leucophore layer, which takes on the color of whatever light shines upon it. This is how a cuttlefish can become a rock, a blade of kelp, a patch of sand.

If you Google "cuttlefish camouflage," one of the first images that appears is of a cuttlefish placed upon a black-and-white checkerboard. The cuttlefish has changed the color of its back to become a checkerboard too, a bright white square against black edges. The more I look at it the less amazed I feel. It feels wrong to ask something as soft and amorphous as a cuttlefish to hide inside the sharp angles of a square. I read about scientists in Massachusetts who connected the wire from an iPod to the fin nerve of a squid to trigger its chromatophores to flash in time with the electrical current of a hip-hop song. They filmed the result, the orange and cranberry sacks of the squid's muscles expanding and contracting in time with "Insane in the Brain," which I suppose is part of the joke. I suppose anything can morph if you force it to.

When it comes to self-defense, cuttlefish are capable of more than camouflage. They can develop distinct disguises for different predators. If a sea bass draws near, the cuttlefish winks with two dark eyespots to appear like a much larger face, perhaps of a much larger animal. If sighted by a dogfish, the cuttlefish blushes dark and flees.

We often conflate the cuttlefish's ability to morph with camouflage, assuming they only use it to hide. Nature documentaries call cuttlefish the masters of disguise. This strikes me as the least interesting thing about the cuttlefish, not just because of the dreariness of the backgrounds they most often blend into but also because camouflage is a body language deployed against predators and others that would harm or devour you. Reading a creature through its camouflage seems a misguided attempt to understand its true nature, its whole self. It would be like studying a zebra while it flees from a lion, or a mouse as it cowers in a hollow log. I want to know how cuttlefish morph when there are no sharks around, only other cuttlefish. I want to know what kinds of transformation the cuttlefish is capable of when it is motivated not by fear but by community and sex, and I am not interested in calling it a disguise.

On one of our first dates, my first girlfriend negs me by asking if she is the last woman I will ever date. By the end of the summer, ghosted and heartsick, I ride the train for hours with no particular destination and watch everyone walk on and off. There are moments when I want to kiss everyone, right there on the train, and there are moments when I cannot imagine wanting anybody else. I am a thumbtack, fallen from a corkboard.

When I return home before senior year of college, I find what seems to be the gayest salon in San Francisco: one named after cum, in the Castro. I show Suzie, my stylist, a picture of the most famous lesbian on Netflix as a reference, and she smiles at me, tucks the photo in her pocket, and never mentions it again, which I later realize is for the best. Suzie washes my hair and asks why the change, and I tell her with a flush that I've just started dating women—maybe for forever, I brag. I recount the banal saga of my first queer relationship, and she listens with unnecessary kindness. When I see the long black strands fall to the floor, I swallow my gasp. But when the razor curves around my neck and scalp and bristles around my ears, I feel electric, like I am buzzing too.

In the years following, I have dreams in which my hair is long again, to my shoulders or my chest, and I have nothing sharp at my disposal, no way of cutting it off. My dream self opens and rifles through drawers, searches for scissors on the sidewalk, loose hair knotted up in a heavy bun. The dream quiets whenever I shave my head again. For some time, this is enough, until it's not.

For a long time, many cuttlefish scientists focused their research on the male cuttlefish—a historically common practice in many fields of science. Among the giant Australian cuttlefish, smaller males change the patterns of their body to appear female, allowing them to evade detection from the kingpin male cuttlefish while sneaking in to fertilize the kingpin's female mate. A PBS nature documentary deemed this "a devious drag act," while a story in *Nature* went so far as to call the cuttlefish a sneaker and a transvestite.

A notable exception to the male bias was a study in 2006, which found that female common cuttlefish share an indicator of recognition called splotch. Splotch looks like it sounds: milky white splotches across the cuttlefish's head, arms, and mantle. The scientists found female cuttlefish only splotched to other females, or to their own image in the mirror. They did not know why the females splotched. Was it a form of communication? If so, what did it say? But the scientists noted that female cuttlefish treat other cuttlefish as potential threats, splaying their arms until they receive a specific signal, perhaps splotch. They suggested splotch could be a way of preventing attacks from other cuttlefish, a physical affirmation of safety and sameness, like the gay nod. I see it as a kind of love language. If you splotch, I splotch.

Male and female common cuttlefish are virtually indistinguishable to humans, which is to say both can change their appearance at will to look like anything they want. The only way to sex a living cuttlefish, according to science, is to place the creatures in front of a mirror and see which flare up in harsh black and white stripes, a display called "intense zebra." Male cuttlefish don intense zebra when they see another male, an overture of aggression and the promise of a fight. But even this is no sure thing, because cuttlefish, like scientists, can be mistaken, and the only way to know for sure is necropsy.

After college I move across the country to Seattle, a city that I thought I would love because my ex loved Seattle and I am still in love with my ex. I can't stop thinking about them, and caught in that old queer quandary of figuring out whether I want to be with someone or be like them, I wonder if it must be the latter. Theirs is the masculinity I know best, used sweaters and functional windbreakers, like I'm a walking L.L.Bean catalog, and I certainly feel queer, but I also feel white, like I am cosplaying as someone who boulders. I start wearing sports bras as a means of disappearing my chest. I am occasionally sir'ed and do not mind. When I start an intensive course of the acne drug Accutane, I am reminded constantly by my doctors and pharmacists that I could get pregnant at any moment, even though I am not having sex, because the drug causes severe birth defects to babies. So I am automatically enrolled in the iPLEDGE Program Guide to Isotretinoin for Female Patients Who Can Get Pregnant, in which each month I have to take a pregnancy test and answer multiple-choice questions about safe sex, about when my partner should put his condom on (answer: as soon as he gets an erection) and whether I can get pregnant underwater (answer: apparently yes). I can't tell if the program is helping me realize that I'm not a woman, or just that I'm not an iPLEDGE woman.

While I live in Seattle, my office is an eight-minute walk from the aquarium. On slow news days, I visit, walking past the black-lit jellyfish, the solitary giant Pacific octopus, the green tentacles outstretched from the touch pools, and sit down next to a small tank holding the dwarf cuttlefish. Their skin flares dark purple in recognition, not of me specifically but of my general presence, which I assume resembles a looming shadow at a worrisome distance from their tank. Flashing is the opposite of camouflage: it transforms the cuttlefish into something unmissable, a mirage of colors that can mesmerize prey, distract a predator, or enchant a mate. Dwarf cuttlefish flash when they want to be seen.

Some mornings, when the aquarium is close to empty, I sit with them long enough that some stop flashing. Their skin prickles, dark clouds and ink blots rippling across it. Cuttlefish have expressive eyes that seem to watch you keenly, their squiggled pupils unreadable and their sheer fins undulating so fast they blur. We watch each other nearly every week for half a year, and I keep hoping vainly that they will recognize me and no longer feel the need to flash their colors in warning. When I visit the aquarium for the last time, six months into Accutane, my own pimpled skin now smoothed out and waxy pink, they flare up and I can hardly blame them—I've been morphing too.

Though cuttlefish may be known best for their camouflage, their visual language may reach its most dazzling and complex when they are speaking with others of their own kind. Splotch and mimicry aside, cuttlefish may create signals that only other cuttlefish can perceive, using a body language invisible to humans. When a sunbeam trickles down into the sea, it is polarized, meaning light waves are made to oscillate in more similar directions in the water. Cuttlefish have some of the most attuned polarization vision known to any animal. Scientists suggest cuttlefish may depend on polarization the way we depend on color to help us perceive our world. Cuttlefish adorn themselves with patterns of polarization, subtly shuffling their reflecting iridophores while their chromatophores and papillae remain unchanged, likely in order to speak to their kind without alerting predators.

Of the cuttlefish patterns we humans can perceive, the most spectacular is called "passing cloud." The pattern is not still but in constant motion, a dark ribbon that moves along the creature's body like a rippling conveyor belt. It is as if the cuttlefish has become a green screen, a portal for other sea creatures to see the sky. The display can resemble the blurred Vs of geese in flight, shadows reflected in a puddle, a psychedelic zebra, as interpretable as a Rorschach test. Scientists have observed passing cloud in swimming cuttlefish, mating cuttlefish, hunting cuttlefish, and resting cuttlefish, which is to say scientists have no idea what it means. Don't eat me/hello there/will you have sex with me/I'm poisonous/that shrimp looks tasty/don't mind me, I'm doing fine. It all depends on the audience.

Cuttlefish can even conjure themselves out of inkblots. Like octopuses and squid, cuttlefish keep an ink sac shrouded under their shimmering skin. Most of the time, it functions as a device of last resort; when the cuttlefish is in danger, it squirts out a smoke screen that veils its escape. But some cuttlefish squirt out a mirror image of themselves, ephemeral silhouettes in ink bubbles and mucus. The viscous decoy, called a pseudomorph, is filled with dopamine, which can stupefy a predator's sense of smell. As the cuttlefish escapes, the predator is stunned, searching for its senses, wondering how a body that seemed so solid could evaporate into ink.

When I move back to New York, I start dating someone in business school. We attend parties with the other handful of queer people in business school, where everyone wears chinos and talks about their summer consulting internships. When we go to a gala thrown by the business school, I half listen to conversations in which people say things like "emerging markets" and "investor relations," and when I go to the bar, business school men tell me I am beautiful. The next weekend I shave my head and get a chest tattoo. I feel my body rebelling, trying to become the opposite of the world she inhabits, and a few months later I realize this, too, is not enough, and we break up in a Café Grumpy.

I start dating someone else soon after, and I spend so much time at their place that I begin wearing their clothes, first as a joke and then by habit. When I go to a bodega in their sweatpants and hoodie, the cashier greets me as if we've met before, and I realize I look just like them, a reflection in a trick mirror. As I walk back to their place, I watch myself in the reflective glass of bodegas and bookstores and feel either lust or self-esteem or a strange mingling of both.

When I return with the fruits, I tell them breathlessly about the incident in the bodega, how I almost passed as them. "Isn't that funny!" I demand, pomelo on my tongue. They laugh and gently remind me that we actually look nothing alike but are simply both Asian, which, for some people, is enough.

When I take the train home, this time in my own clothes, I wonder if I have developed a pattern of becoming subsumed by whomever I want to fuck, my appearance morphing into a reflection of my crush or my lover, or a refraction of my ex. I find myself wearing clothes my exes gave to me or unknowingly left behind: a pair of shorts, a Uniqlo button-down, a green puffer jacket. Sometime later, lonely again, I get an enormous tattoo of a foo dog on my thigh. I tell my friends I got the tattoo to feel connected to my Chinese heritage, which is half-true, and when I come home and peel off the Saran Wrap, the ink has corrugated my skin into a dense topography. *Tattoos don't have to have meaning,* I tell myself as I empty a travel-sized tube of Aquaphor onto my burning leg. *It's okay to get a tattoo to feel hot.* But I know I did not get this tattoo to feel hot; I got this tattoo to be desired, and I begin to wonder how much I have changed my body for me and how much I've changed it for others.

Flamboyant cuttlefish get their name from their jeweled coloring, flashing magenta and gold. In an aquarium, they are a riot of color, pulsing and shimmering and waving their small mouthful of tentacles at humans who draw too near. But in the wild, flamboyant cuttlefish are rarely flamboyant, unless they're scared or horny. The cephalopods spend their days trundling through mud and looking like mud. But if a predator approaches, the cuttlefish may burst into color to confuse its would-be devourer and escape. The flamboyant cuttlefish is only flamboyant around bright lights and looming shadows because it is startled, flushed, endangered.

The other exception is when the flamboyant cuttlefish decides to mate. The small male splashes himself in passing cloud, his skin a tapestry of regal reds, white swirls, and a long gold stripe. He dazzles, rippling chiaroscuro to the camouflaged object of his affections. This courtship can expose the cuttlefish to passing predators, even land the pursuer in the jaws of a scorpion fish. But if he's lucky, the female cuttlefish will concede to mate with him, and she will lay her eggs in coconut shells, the only safe places on the open, sandy seafloor.

Flamboyant cuttlefish can live up to two years and mate in the spring, over a handful of weeks during which each female mates with as many other cuttlefish as she desires. She will die soon after she has laid her eggs, so she makes the most of it while she's still soft, pulsing, and alive.

When I am in college, my friend Evan and I take what we later suspect is synthetic acid. The trip starts out normally enough: we skitter in and out of time like water striders and give in to a fevered urge to merge with the grass, melt into the earth. But when we go back to Evan's apartment, I see the walls caving in and doorways collapsing. I see the glassy brown marble of Evan's eye drip down their face until it dangles from the sharp line of their jaw. My mouth opens in horror and my hands reach out instinctively to catch it before it falls to the floor. I watch my own arm bend in a perfect, taffied curve as if I had no bones, nothing rigid inside myself to support all this flesh and blood. Our friend Liz holds the limp tube of my arm, squeezes it, and tells me I am okay, that my arm is not broken. I take a shower to collect myself, and I try to look at the tiles, but they are changing color. I focus on the grout, but it is creeping out of my vision like a slime mold. I look at the cartoon eye on the L'Oréal shampoo bottle, which drips off too. But I can't help it and I look at my body. It is wobbling in front of my eyes: legs wavy like bacon and one breast snaking toward the floor. The other is nowhere to be seen, just a nipple lying nonchalantly on a flat chest like an eraser on a desk. When I get out of the shower to tell Evan about this sensation, we agree it's impossible that I could have seen my body shape-shift. Denying it helps calm me down. But what I felt was real—the rubbery blade of my arm, my fingers sprouting like seedlings, the tingling cosmos of my infinite chest.

I block this trip out of my mind for years, so close to

forgetting, until those weeks this past summer when I am staring down at myself in the shower, distant feet and pale legs, and suddenly I feel something, a pang of remembrance like a phantom limb, and hold my fingers up to see if they will shrink and grow. They do not. I have missed my chance and am now stuck in my immutable body, my cast.

Unlike chameleons, cuttlefish can change textures, becoming as smooth as a marble in one instant and, in the next, fringed like seaweed. The change comes from their papillae, bundles of muscle that can expand from smooth to jutting, somewhat akin to a hardening nipple. They have two sets of muscles, and alternating which ones they squeeze and relax determines whether bumps arise or retreat back into flesh. One scientist compares it to squeezing a water balloon—cinch it in the middle and the top will rise; let go and it all sags back.

Cuttlefish can mime texture on sight alone, never needing to physically touch the substance they imitate. Watching a cuttlefish's papillae extrude and hold stiff feels almost tectonic, like you are witnessing the birth of a volcano or the growth of a forest from a bird's eye. Each cuttlefish species has a repertoire of various papillae shapes—spikes and spines and mushroomy blobs—allowing the creatures to smooth out on a bed of pebbles or prickle against branches of coral, as though the surface of their skin is dancing and alive. Once the papillae are shaped as the cuttlefish desires, it locks them into place so it can relax the rest of its muscles and swim about with ease. Cuttlefish can keep this up for an hour without expending any energy, their body locked in this new transfiguration, their senses free to experience other things.

This summer, what starts as a longing to please my partner while holding their face in my hands soon morphs into a longing for a cock, not just something I wear but something that could feel, occasionally, like it was a part of me. Until this point, the strap-ons I have worn have belonged not to me but to my partners. The first was a lavender dildo slipped into a specially designed pair of cotton briefs— very comfortable, but we could use it only once every wash cycle, which, in college, was infrequent. The second was something eerily vintage, a musty gray pillar that wobbled out from a harness seemingly made of briefcase leather, all buckles and straps and cracked around the edges. The third was some matte-black cigar blooming from a linty Velcroed fabric, and the fourth, grooveless and aubergine, as so many of them are.

I begin to think about a strap-on outside of the realm of sex, how nice it might feel to walk around and see something hard and embodied by my groin, and so the first strap-on I buy is for myself, though I eventually use it with my partner, T. I put off the decision for months, torn between the options and unsure if anything I've book-marked will give me a wedgie, will wash easily, will sit comfortably on my hips. I spend hours scrolling through reviews, dizzied by the options. I marvel at how mine can be any color I want: purple, black, red, teal. Or even prettier: amethyst, obsidian, tourmaline, lapis lazuli. It can be glittery, metallic, tie-dyed with clouds. I wonder if it should be textured, eerily veined, or more abstract—grooved like a wrung-out towel or ribbed like a squash. It could take on any shape it wanted, bent at the tip like a finger, or bulbous like a mushroom. I see one shaped like the inside of a geode, a long periwinkle crystal. I can't imagine any softness to it; it looks like it could cut me open. But when I hold it in person at a Babeland, it bends, willingly, in my palm, its jaggedness an optical illusion.

I do not buy the geode but something simpler, black silicone and a tangle of silver hardware and black leather emblazoned with a jaguar staring out above a metal ring. When it arrives, it smells like a handbag and feels like butter. That night I put it on and walk around alone in my room, feeling the heft of my cock bob up and down. I've tried it on for size but can't seem, or don't want, to take it off now. When I close my eyes, it feels almost like a part of me: a temporary extension of my body, something that has always lurked but only now emerges, erect and pliant. When I finally take it off, I place it next to me and stroke it.

When I squeeze it, it seems to grow. It is an out-of-body experience, both a part of me and a part, abstractly, of my partner. Having it changes something, puts me at ease. I feel it is mine when I am not wearing it, when it is in a drawer, or when it is on T. I begin to think of my body as capable of being disassembled, scattered, reunited.

The scientific name for cuttlefish is *Sepia,* and the cuttlefish gave the color its name, not the other way around. The ancient Greeks used cuttlefish ink to write with, stabbing their nibs in dead cuttlefish's ink sacs, producing the distinctive, almost translucent brown hue. Sepia has come to describe a color rather than a substance, and it is a color associated with the past, antique films and photographs all beige and discolored. When a cuttlefish's ink is taken from its body, the resulting drawing or letter will always appear vintage and outdated, no matter how freshly drawn.

Each time I try to write this piece I feel differently about my body, my gender, myself. Each time I conclude that I must not be ready to write it; best to experience the thing and then wait a few years to reflect, the advice generally goes. But if I don't write it now, how will I trace my own evolution? So I dub this essay a pseudomorph, a gibbous moon, a silhouette in ink of the person I am now and whom I may no longer resemble in the future.

In the past two years, T and I have phased out of calling each other partners (overly official; conjures corporate imagery) and into calling each other boyfriends (fun and approachable). It started, I think, after passing someone on the street who mistook us for boys, which we soon agreed was not really a mistake after all. This has led to rare moments that feel like flashbacks to a past, heterosexual self—when the movers ask if I'm moving in with my boyfriend (yes, I say, both of us seeing the situation in different lights). I haven't said "boyfriend" this much since I thought I was straight, and the word rushes seamlessly back into my vocabulary with all the glow of nostalgia but none of the baggage of heterosexuality. We keep our cocks in a drawer in the bedroom, each bagged in a ziplock to keep off the dust, next to clothes I no longer wear but cannot bear to give away, each a self I could embody in my brief, unguarded life.

In these last few months, I've started showering with the lights off. The first few times I slipped on wet tile, squirted shampoo somewhere that was not my hand, fumbled for the knob when the water grew too hot. Now I leave the door ajar to let the light in, just enough not to slip. I turn the knob and step into the warmth crackling from the spigot. The water streams down me in rivulets, and I imagine it enveloping me in a bubble, or rather the opposite of a bubble—a pocket of water surrounded by air. I close my eyes and imagine my fingers sprouting, my jawline widening into a shelf, a deep V tunneling down my hips. I imagine a body that is different every time. Sometimes it is unattainable: enormous biceps and a rippling back. Sometimes it is within reach: scars and sewn-on nipples. These are times when I am thankful I don't live in California anymore, able to let the spigot flow without thinking every second of drought, able to imagine myself in water: refracted and scattered until I am something like light.

Us Everlasting

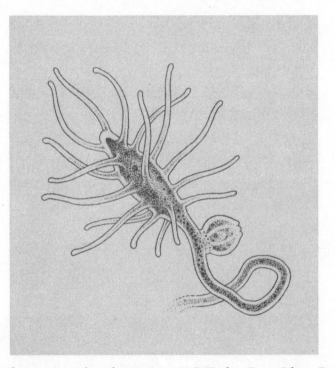

Contributors (in order of appearance): C Taylor, Evan Silver, Futaba Shioda, CV Sise, Rosemary Hartunian Alumbaugh, Alexis Aceves Garcia, Zefyr Lisowski, T Zhang, Amy Jackson, Kirstin Milks, Camille Beredjick, Gabriel Stein-Bodenheimer, Darcy Curwen, and Martha Harbison

In the ravenous ocean, very few immortal jellyfish live up to their name. They could, in theory, go on forever without dying, but most do not. They do not feel pain the same way we do, may not—even as they are being torn apart—comprehend what pain is, but they can be damaged, and often are. They can age, can succumb to infection or be stabbed into a pulp with tweezers, can be whisked into the bladed fan of a nuclear power plant. They can be swallowed whole. What does immortality mean when you can still be eaten by anything with a mouth? Most immortal jellyfish die in this thoroughly regular way: fortunate enough not to glitch into disease or senescence, unfortunate enough to disintegrate in stomach acid.

It resembles a ghostly tassel, translucent dome topping tentacled fringe. It is tiny, approximately the size of a pea, and when it floats in water, its glassy body winks in and out of visibility as if able to dip parts of itself into other dimensions. It has color only in its gonads—four rusty dots suspended in the goo of its umbrella. The immortal jellyfish usually has sixteen tentacles or fewer, each swelled at the tip like a sticky hand. It has no brain, no heart, no eyes. It has only one hole, a mouth-cum-anus, which eats and excretes and squirts water to propel its bell-shaped body.

The immortal jellyfish starts out life just like any other jellyfish: a fertilized egg morphing into a pellet-shaped larva that settles on the seafloor and unfurls into a branching stalk called a polyp. Like a fruit tree, the polyp's arms effloresce in buds that shed tiny medusas, the final form of a jellyfish. Within a month, the small medusas balloon into a crowd of adults. From here on out, other jellyfish might mature, lay their eggs or spew their sperm, and slowly trickle into death.

But when the immortal jellyfish's body begins to fail, it ages backwards. Its ailing body sinks to the seafloor or some hard surface and rearranges itself into a silky lump, looking like an egg, or a cell, primordial, all potential. It seals itself in an envelope of chitin and shuffles around the meaning of its cells. And then it sprouts into a polyp and grows, into not one individual jellyfish but many clones. So the single damaged jellyfish becomes a host of younger, possible selves, each with the same power to regenerate. As far as we know, the jellyfish can do this over and over and over, as many times as they are damaged, into something like eternity.

The scientists who originally described the peculiar regeneration did not choose to call the creature, also known as *Turritopsis dohrnii,* "immortal." One scientist went so far as to write that he would never have used that word; he originally said the jellyfish was capable of "ontogeny reversal," meaning reversing the direction of the ordinary life cycle. But "ontogeny reversal jellyfish" does not make headlines or earn grant funding, and so the creature became immortal. The jellyfish does not know it is immortal, nor would it long for such a thing if it were not. Jellyfish do not long for anything. They never feel out of place, because they do not feel. Some scientists say this biological simplicity may explain the jellyfish's immortality, a kind of evolutionary trade-off. When I read this, I understand: if something intelligent were ever given a second chance at life, it may never want to grow up.

Like millions of other mortal beings, when I learned about the immortal jellyfish, I envied the animal. It wasn't the immortality itself that I coveted but its mechanism. Our traditional notions of immortality are so languorous and passive: Jesse Tuck, the everlasting teen, sips from a magic spring and stays seventeen for eternity, just like Edward Cullen from *Twilight.* I grew up thinking of immortality as something won with a drink or a bite or a pill, a static and irreversible state of being. But the immortal jellyfish has no notion of these tepid forevers. Its immortality is active. It is constantly aging in both directions, always reinventing itself, bell shrinking and expanding, tentacles retreating into flesh and wriggling out again. It is not living forever but reliving forever. When the immortal jellyfish ages in reverse, its body is not choosing eternity but rejecting death, which seem to me entirely different things. And the life it chooses is not young adulthood, forever a gangly seventeen, but childhood. It grows up again.

It's an old trope now that many queer and trans people have a second adolescence. The first happens alongside everyone else's, except you are not yourself. You feel as if you are the only person you can talk to. You live someone else's truth. Or maybe you don't know how to be a queer child, never even considered it for one or every reason, and things don't make sense until one day you turn twenty and have what some call a breakthrough and the clips from your childhood roll back in a suddenly sense-making montage. Watching Leonardo DiCaprio gradually freeze to death on loop and wondering if he was the only man you would ever want to kiss. Sleeping next to a three-foot-tall poster of Shania Twain on the wall. Dreaming of girls in your class and waking up in a fog of confusion, or shame. Now comes your second childhood, second adolescence. Maybe you cut off your hair, or maybe you wear chokers. Maybe you fall in queer love for the first time, which feels symphonic when it starts, world-ending when it's over. This second adolescence is bittersweet — full of highs and also plagued by the nagging reminder that all this could have been yours the first time around.

So what if you could do it over? And then again? What body would you choose? Who would you be and who would you love? Would you do it over, and over, and over again?

Life #2: C

I am twelve or maybe thirteen, the age at which I start to borrow my mother's electric razor on the sly. The age at which the meanest girls on my soccer team are whispering about my leg hair, the dark spikes above my shin guards. They whisper loud enough for me to hear. Soft enough that the coach, gay and herself the subject of whispers, does not.

I do not shave at twelve. I do not shave at thirteen. I do not shave at fourteen or fifteen or sixteen. I do not shave. My tender-flesh thighs never redden or itch. I never cut myself by accident in a sudden sting, the blood pinking down my ankle in the shower. The hair grows. The hair grows, and silkens, and covers me whole. The hair catches the wind on my long, bare-legged bike rides. I imagine sometimes the hair furs me inside too, from lung to liver. I am all cilia and no shame. I creature through the Midwest winters: hirsute, werequeer, warm and soft to the touch.

Life #3: Evan

At Camp Timberlane for Boys, Jason and I survive by practicing spells in the woods with gnarled twig wands. Eventually the other kids take notice. They stop playing touch football and capture the flag and instead assemble to watch us practice witchcraft. There is no need to hide our powers from others, nor to repress them so deeply that it would take over a decade for us to realize that we were magic all along. When Jason's profile shows up on a gay dating app many years later, I am not flush with shame and remorse. This time, I say hello, and ask: "How is life?"

Lives #4–8: Futaba

When I raid my father's closet in middle school, I take *all* of the ties. I learn how to tie them properly and wear them every single day of the seventh grade.

At recess, when the boys won't let me play football, as the teachers watch and laugh, I deflate all the footballs and bury them in the ground.

In the hotel room, after the pretty girl does my face with makeup, I vomit in the toilet. I run out into the hall, screaming. When I ride the elevator, I cannot see my face in the ceiling mirror. I have no reflection, and I am floating a million floors away.

If I could go back to that relationship, I would pay more attention: how she wanted me to grow out my hair, how I wanted her to grow out her beard. Maybe this should have signified to us that we wanted to switch. Maybe we were in love from the other sides of the looking glass.

Let me tell you again how the immortal jellyfish does it, so you can try it too. You begin as an adult, nearly invisible, with all your tentacles. The edges of your umbrella brim with buds that can unspool into new tentacles, if needed. The domed bell of your body is lined with hollow pipes sprouting from the bottom of the tube that contains your mouth. The pipes carry food throughout you, all entering and exiting through your one and only hole. Your umbrella is tiny, no larger than a sesame seed. With a body 95 percent water, you are more like a lake than an animal.

You pulse through the ocean aimlessly — jellyfish have less need for direction — until something happens. You are stressed, injured, caught off guard. Maybe the chemicals in the water surrounding you have shifted. Maybe a part of your body has gone missing, liberated by the jaws of something larger.

So you transform. Your mouth tube shrinks and your buds retreat back into the edges of your bell. Your tentacles disintegrate into shreds and stumps. Your saucer-shaped body pinches itself into a four-leaf clover as if wishing yourself luck in this new life. Your smooth edges disappear into cracks and scalloped edges. Parts of you have become unrecognizable to yourself.

All this self-destruction makes you shrink, and you are now a quarter of the size you once were, smaller even than a poppy seed. The scaffolding of pipes that snaked through your body now fuses together. You shed whatever muscle once clung beneath your bell. Your body shape is now unmistakably ovoid, small pill of you. Now you encase yourself in an outer layer and shape-shift, slightly, from oval to ball. There, the hard part's done now, so you let your tiny round body cling to whatever substrate is closest—rock, coral, glass, anything will do—and rest. You can take as long as you need.

And now the most mysterious part, the stage we still don't quite understand. You begin to transform the very nature of your body. You take a wrecking ball to the cells that once distinguished you as an adult with a mature body and reshape them into cells that serve the next iteration of yourself. All your potpourri of nerve cells and ectodermal cells and endodermal cells transform into stolon tissue, building the branchlike body into which you will soon grow. One type of adult cell metamorphosing into another might seem impossible; only stem cells can become anything they want, the rule goes. But somehow you've fashioned your own mechanism of reinvention, made your old cells new again, given yourself whatever you need.

When you are ready, your body prickles into tips that break through and branch out. You begin to resemble some kind of gelatinous palm tree, sprouting from your perch and unfurling your tentacles like a crown of leaves. You are a polyp now, rooted firmly, and you feed on anything that drifts by, your tentacles snaking wayward flakes of flesh into your mouth. Soon you will sprout your first bud, which whirls away as a tiny bell-shaped medusa, a younger clone of what you once were. You bud and birth more and more selves, each juvenile jellyfish sharing your past but determining its own future. All you, and you, and you, and you, and you. All your possible futures, each bobbing away in new directions. Your old body is gone now, but who could think of the past when you are now these young and perfect jellyfish, all of you whole again, ready to eat and drift and live and hurt in different ways than before.

Life #9: CV

In college, I work in a cadaver lab. Each day I take human limbs out of beer coolers where they are labeled in Sharpie with blunt descriptors: "Arms" or "Ankles." My job is to thaw and prepare them for dissection, to take note of their IDs, and to clean up post–medical lessons. Sometimes the lab manager lets me dissect the limbs and the surgeon's handiwork after the med students leave. I tug on tendons like pulleys and watch the toes, fingers, and calves move in response—what a limb becomes when it's apart from the body.

When dissections are done, my last job in the cadaver lab is to package up the limbs for disposal. I tuck them neatly into waste bags and put them in a small metal bin, a trash can for hazardous waste. Looking at a hand propped up by the circumference of the bin, I wonder if after this final stage of disposal it will be reunited with its owner. Sometimes I dream about the limbs in buckets. I wonder if something spiritual can grow in place of something physical. What happens to the whole with a part removed?

One night in the cadaver lab I decide to flatten my chest. I place it on the medical table and feel as I did toward my other dissections, a distanced sensation of awe and respect for the human form. I write out my own sample ID:

CV, 20 years old

and hold its soft shape for the last time. I carefully study the skin pressed sore from the binder and strip away each gland, forfeiting function. I wrap up the flesh in a hazardous waste bag and throw it out.

Life #10: Rosie

I tell my seventeen-year-old self there will be more worship in kissing women than she'll ever find singing about sin. I think she already knew. But I want her to have the freedom I never had, so she believes it. So she does not spend so many futile hours crafting biblical arguments to convince religious leaders of her holy and sacred queerness. So she does not put her physical body through hell that summer. She buys her first swim shorts five years early. She kisses her Bible study leader behind the chapel during the Altar Call.

Life #11: Alexis

In 2021, I take a hypnosis class to visit myself in the past, present, and future. As I focus on the words and the flame of a lit candle on my coffee table, I open my eyes between two rows of lemon and mandarin trees in my abuela's backyard. I float up into a trail of golden light above me, unsure of where I am headed in my own timeline.

I visit an overworked version of myself in 2018 walking to work on West 24th and Fifth Avenue in Manhattan: steamed shirt and depleted optimism. I yell tender things at them. Something like, *"I love you and you're on the right track. I know nothing makes sense and it won't really in the future either, but you'll find moments of deep joy and closeness to yourself and I love you, I love you, I love you."*

I slide into a picnic table next to myself during a hard conversation about gender with my mom at a burger restaurant down the street from my apartment. I am there to be a witness. To sit next to myself as multiplicitous as I feel day-to-day.

I visit with myself in preschool, when I spoke only Spanish and didn't have many friends. Little Alexis gives me a tour of the classroom. Points to the world map on the back of a chalkboard on wheels, the laminated English alphabet peeling from the wall above it. And it's here I wish I could regrow from. Plant defiance in my hand like a fragrant pink rose from Abuela's driveway. Grow a NO so large in my body. Allow it to ripple through timelines and versions of myself. I had a trans childhood because it was mine and it's not over. Each time I learn to protect myself now, I protect myself then. I make space for new versions of myself in the future. A forever expansion and re-creation of my own life.

*　　*　　*

Of course there's a catch. There's always a catch. An immortal jellyfish can't go back whenever it wants, only when it has no other choice. Healthy immortal jellyfish, presumably content with their tentacled curtain, cannot flip the chemical signal to age in reverse at any moment of their choosing. Trauma is not just a catalyst to regeneration; it is the only catalyst.

Scientists who study the regeneration of the immortal jellyfish know this well, and have developed a roster of abuses to "induce rejuvenation," as one study calls it. One standard method of traumatizing the jellyfish is to place the creature in a solution of cesium chloride, a colorless salt. An alternative, called the needle treatment, asks you to pierce the gooey umbrella with a stainless-steel needle. Some scientists drag the needle through the creature's body in a scribble, removing the needle as burst cells coagulate like cumulus. Others stab repeatedly, up to fifty times per jellyfish. You can also heat shock the jellyfish, raising the temperature of the surrounding water to nearly 100 degrees. Or you can simply starve it.

If you do not give the jellyfish more than it can handle, it will not begin to regrow. If there is not enough cesium chloride in the petri dish, if there are not enough needles or there is too little heat, the jellyfish will remain adult, alive. So you have to ensure there is enough stress, enough trauma.

One scientist in Japan, Dr. Shin Kubota, has kept a group of immortal jellyfish regrowing in an eternal loop, growing old and growing young, as many as ten times in two years. The jellyfish, which descend from wild jellyfish Dr. Kubota collected off the shores of Okinawa, live in otherwise ideal conditions: in a vessel with a steady flow of seawater somewhere away from the sun. Dr. Kubota, who has devoted his life to studying these jellyfish, is likely the world's leading expert on keeping them alive regeneration after regeneration. Some immortal jellyfish have died in his care, not from needles but from algal growth.

There are no studies observing whether, and how, the immortal jellyfish regenerates in the wild. The creatures are too small, too close to invisible, too hard to find. But there are also no studies observing how the immortal jellyfish might regenerate on its own terms in the lab. Scientists only want to study the creature if they can poke it back into polyphood. It would be too time-consuming, too expensive, too thoroughly unimportant to the interests of humans to examine the jellyfish without our interference. So we thrust the few living in our care into looping lives. They live and relive and relive not due to any natural trauma or the danger of the oceans but because we want to see them age backwards and we have the tools to make them do it. We have made it impossible for the jellyfish to continue on in their second, third, fifth, seventh adulthoods because we want to see it over again. And yet each time, after each salt bath or stabbing, the jellyfish come back, their bodies retracing an ancient blueprint. They just can't help it.

Life #12: Zefyr

I'm twelve when I learn to fight, rolled-up carpet hanging from the garage a makeshift punching bag. My father is the one teaching me, skin sticky in the Southern heat; it's 4 p.m. so he isn't drunk yet. It takes me years to hit back, and when I do, the boy at school crumples like an aluminum tray, but before then, I dream of a different childhood. Different me has a more queer haircut and no regrets. She spends her teen years chopping things apart. She chops the furniture in the ranch house into tinier pieces of furniture. She chops my father into a kinder father.

What does violence turn into if I could go back and shrink it? My skin de-bruises. My muscles never know the language of assault. Our recycling bin is now empty of my father's boxed Franzia. I'm sitting on the grassy banks of the Pasquotank with every friend I once loved, and maybe our hands are touching—or will soon. What's left of queerness when it's not defined by violence endured? The carpet is on the floor of the living room. Reflecting off the muddy water, our skin gleams like hope.

Life #13: T

José Muñoz says queerness is an ideality. A place and time that exist beyond this one. It's a beautiful image, a yearning that reaches toward a horizon. Like so many beautiful things, it's also incredibly sad. Queerness for Muñoz is a promise that's always just out of reach. If we can live again and again,

I wonder if Muñoz's formulation transforms into something else entirely. Instead of existing nowhere, queerness might exist everywhere at once, in an infinite multiplicity of nows.

Would you be surprised to learn the immortal jellyfish is taking over the world? And it's our fault—at least if you consider it a bad thing. They've breached oceans moored to the ballasts of our ships, and now they blink their tentacles in waters off Panama, Japan, Italy, Spain, New Zealand, Tasmania, and even Florida, the place where strange creatures go to thrive. We only noticed their presence because we looked for them. We sifted through waters in search of something clearer than the murk, glass thimbles glinting in and out of sight. And once we documented their presence in these far-flung places, we declared it a silent invasion: silent because we never noticed it before, an invasion because the jellyfish had arrived somewhere we did not expect them to be. Maybe we call them invaders because we are jealous of them.

The truth is that many creatures are capable of renewing themselves. Newts can regrow their limbs, sea stars their arms. Zebra fish can regenerate their fins, their spinal cord, their retina, and most of their heart. This isn't immortality but a second chance. Sometimes you don't need to pull yourself apart to start over. Metamorphosis isn't always a full-body thing.

Since we first observed the immortal jellyfish's capacity for resurrection in 1988, other jellyfish have revealed variations on this regenerative ability. In Italy, a translucent jellyfish shaped like a miniature space helmet revealed its second life in a lab. Scientists had collected the jellyfish, called *Laodicea undulata,* with the sole intention of reconstructing its life cycle: how it is born, lives, and dies. In the lab, the baby *Laodicea* sprung forth from the polyps and lived for less than two weeks before twinkling into death, except for one. This last *Laodicea* had lost all its tentacles but two. It sunk to the bottom of the tank and within hours had transformed itself into a ball. And then the familiar steps began: prickling of stolons, one polyp bud followed by many, and finally three young medusas, bobbing in the same tank.

In China, a graduate student unwilling to part with a dead moon jellyfish, his companion for more than a year, collected fragments of its corpse and placed them in a new tank. More than two months later, he found a polyp with three tentacle nubs sprouting from the jellyfish corpse. More and more polyps unfurled in the following days, and the graduate student dutifully collected them and placed them in a new tank, where they settled in and fruited and bloomed into medusas. Moon jellyfish are also often invaders, blown by currents and rippling into vast blooms in the Atlantic and the Pacific. They are one of the best-studied jellyfish in the world, and yet no one noticed the moon jelly's power of regeneration until someone gave it time and trust that it might grow into itself. Perhaps any jellyfish is capable of such transformation.

So I have to ask you again. How shall you regrow, and in how many ways?

Lives #14–19: Amy, Kirstin, Camille, Gabe, Darcy, Martha
…What I think of as a "dream childhood" for me isn't particularly fanciful. I'm still trans in it, because I cannot imagine the world without that fact, but I just come to that conclusion much, much sooner and I'm allowed to explore it. What happens if I give my eight-year-old self a dress?…

…I go back to the first time I heard someone I loved use the word "wife" in pillow talk. My rib cage responds as it notices how my heart yearned to disappear, pulling me from the bed, no matter how strongly my arms thought they should hang on….

…Today, it makes me so sad to think about how much more I could have loved myself, and how much sooner….

…I don't walk back home the morning after. Don't kick a can down the three blocks. Instead, I wake up in her bed and savor the snippets of memory: the dregs of the seder Manischewitz, her head on my chest while we talk and sprawl on the kitchen floor, the wildness of her naked body pressed against mine. No one at school makes fun of us, and in a couple months, when my T kicks in, I'll look like any other boy….

…I give myself a space to breathe. A space to be unseen. I am in a tide pool, in the sun's warmth, and I am floating in seaweed. Here is where I feel safe….

...I grow up inhabiting my body instead of feeling like a ghost driving a stolen meat suit—my little tentacles stretching, stretching out into the neglected spaces of myself, grabbing hold, finally making myself whole....

On the southern coast of South Korea, an autonomous robot named JEROS—short for Jellyfish Elimination RObotic Swarm—is tasked with thinning out the newfound abundance of jellyfish blooms. The robots float on the surface of the water like tripods on Jet Skis, each equipped with cameras, a GPS, and a latticed grid of wires under the surface. The camera and GPS give JEROS the ability to scan the sea for its target, and when it detects jellies, it corrals the swarm like orcas on the hunt. It is fortunate that jellyfish can feel no pain, because once they are cornered, JEROS sucks them in and pulverizes the creatures with thin wires designed to slice through jelly flesh. The robots can shred almost a ton of jellyfish per hour, showering the seafloor with gelatinous mulch. JEROS, the scientists say, is an efficient way to remove the masses of jellies that clog nuclear reactor plants, deplete commercial fish populations by devouring fish eggs and plankton, and swarm, unwanted, in the ocean. It is the best defense they have developed against the jelly nuisance.

But what they do not know, writes the jelly biologist Rebecca Helm, is that shredding certain jellyfish, such as stinging sea nettles, will only trigger their spawning. What is intended as a jellyfish massacre becomes a disembodied orgy, a cyclone of jelly eggs and jelly sperm unleashed all at once. All this spunk meets and refigures itself into embryos that sink and land and sprout into polyps that can produce hundreds of clones, each clone able to produce hundreds of jellies.

What they do not know is that certain jellyfish, despite their name, are unexpectedly tough. The gargantuan Nomura's jelly, which can grow to two yards in diameter and weigh as much as a lion, has skin so sturdy it will bounce off a cutting blade. It is impervious to the shredding robot and the intake screen of a power plant. It is essentially unpulpable.

What they do not know is that tiny jellies sometimes called immortals may, if they are confettied by the robotic blades, sink to the seafloor, the fragments of their bodies coating the bay like gelatinous shag carpet, and begin again. Not everyone will make it, but the ones that do will seal themselves in chitin. Will await the nubbly prick of something like a root. Will rise into a polyp, and then many more, until, some months later, the pearly caps of baby medusas burst free and pinwheel toward the sky. From above, the water might look filled with tapioca. All across the bay, as far as the eye can see: baby medusas, tender haloed fringe and translucent clovers, all rising like fallen blossoms reuniting with a tree. Summer, happening in reverse. All of us moving toward life. All of us refusing to die.

Acknowledgments

This book takes its name from the vertical zones of the ocean, which are divided based on how far light reaches. The photic, or sunlight, zone occupies the top two hundred meters, where waters are illuminated enough for organisms to make food out of light. Below, the twilight zone, which descends to about a thousand meters, is diffused by an almost imperceptible amount of light, far too little for photosynthesis. And finally there is the aphotic, or midnight, zone, a region of perpetual darkness that comprises 90 percent of the ocean.

If a book is an ocean, the final text is a bit like the sunlight zone—a tiny fraction of the research and love and people who contributed to it. So I want to shine the headlights of a metaphorical ROV into the friendly abyss of everyone who made this book possible.

I am immensely grateful for all the scientists whose work sheds light on these creatures and their many ways of living. Several are named in the source notes, but I wanted to give special thanks to researchers whose work directly inspired these essays. Bruce Robison, Brad Seibel, and Jeffrey Drazen's *PLoS ONE* paper on the fifty-three-month brooding period of a *Graneledone boreopacifica* was the first scientific paper I ever read that cracked open my heart a little and pushed me to think about how I could write that connection. Craig R. Smith's extensive research on whale falls reveals the way death can transform not just the dead but the living. Kim Martini asked the research community to dispense with a problematic name for an iconic worm and stand with survivors of sexual violence. Roger Hanlon's close studies of cuttlefish decode one of the most magical metamorphoses in the ocean. And Shin Kubota's infatuation with the immortal jellyfish has helped us mortals become infatuated too.

I would not have been able to access most of these papers without the work of Alexandra Elbakyan, who created Sci-Hub, a site that subverts hefty journal fees. I hope the academic world moves toward open access for all.

And I wouldn't have known to search for much of this research were it not for the science writers and journalists who first introduced me to some of the creatures in this book. It is a gift to read work that captures some of the wonder of the natural world and digests it for people with no formal background in science, like me. I am thankful for Feini Yin's story on the feral goldfish, Jason Bittel's story about the Chinese sturgeon and yeti crabs, Lynda V. Mapes's weeks-long dispatch on Tahlequah's tour of grief, and Juli Berwald's feature on the possibilities of the immortal jellyfish. A cetacean-size thank-you to Ed Yong for writing about the octopus and the whales and the yeti crabs and the salps and the cuttlefish, and for all your kindness.

I am in awe of the generosity and imagination of everyone whose words appear in the essay "Us Everlasting." Thank you for responding to a random email or strange Twitter callout and sharing such dreams with me: Zefyr Lisowski, CV Sise, Evan Silver, Alexis Aceves Garcia, C Taylor, Futaba Shioda, Rosie Hartunian Alumbaugh, T Zhang, Amy Jackson, Gabriel Stein-Bodenheimer, Kirstin Milks, Camille Beredjick, Darcy Curwen, and Martha Harbison.

I would never have dreamed of this book without Megha Majumdar, my first editor at Catapult, who picked up my pitch for a column from the Submittable void and nurtured this collection in its most larval stages. Megha's careful and expert eye helped the column bloom from one essay about a purple octopus into a range of varyingly successful attempts—essays in the truest sense of the word. Without Megha and Catapult, I would never have thought these pieces could be something more.

To everyone who has read parts of the book and offered feedback, input, emotions, and wisdom: I deeply appreciate you and your time. A special thanks to Eleanor Cummins and Marion Renault of the Iowa Writers' Workshop group chat (not affiliated

Acknowledgments

with the official Iowa Writers' Workshop), my Tin House Workshop under the brilliant Cyrus Simonoff, and my Catapult workshop with the luminous Meredith Talusan.

I am indebted to all the organizations that have supported me and this project over the past few years. To Kima Jones, Dr. LaToya Watkins, Allison Conner, and the fellows of the Jack Jones Literary Arts Retreat: it was an honor to share such a magical space with you and to witness your work. To Alexander Chee: thank you for funding the Yi Dae Up Fellowship and your generosity to so many of us. It was an honor to write under your grandmother's legacy. To Jyothi Natarajan, Yasmin Adele Majeed, and the Asian American Writers' Workshop, especially Margins fellows Abigail Savitch-Lew, Amanda Ajamfar, and Yuxi Lin: thank you for all the care and belief and community. To Millay Arts, Andrea Pérez Bessin, Annie Liontas, Dāshaun Washington, Lucas Baisch, and Peg Harrigan: it was a joy to write alongside all of you and take in the spring together. Thank you to Ryan Davenport at Paragraph NY and Christopher Deputato at the Café Royal Cultural Foundation.

I wrote most of this book at night or on weekends while working full-time journalism jobs. (Note: I do *not* recommend this!) But I am grateful for the emotional support of my editors and colleagues. A special thank-you to Daniel A. Gross, Emily Anthes, Alan Burdick, Katherine J. Wu, Arya Sundaram, and Giulia Heyward for kindness and support during the book's final stretch.

I would cross oceans for Hannah Seo, a poet and fact-checker who caught so many small and big errors in earlier drafts. Hannah asked questions that expanded and enriched the book in ways I couldn't have imagined, and their poetic sensibility was invaluable in figuring out how far metaphors could stretch without misrepresenting the science.

I could not have imagined illustrations as stunning as the ones created for this book by the artist Simon Ban (please check out his tattoos at @squids.ink on Instagram). Thank you, Simon, for rendering these creatures so faithfully and with such care and grace. Your art brings the book alive in a way words cannot.

Acknowledgments

Thank you to everyone whose support is less easily categorized but no less deeply felt. Mimi and Neaka for nearly two decades of friendship. Sean for our trips. Luca and Meimei for those beautiful weeks. Love, love, and more love to Elaine Hsieh Chou and Shayla Lawz, dear friends and incandescent writers who have been pillars of support, wisdom, and gossip.

To my friends and family who appear in the book: thank you for the light you have brought to my life and for the gift of your permission to render you on these pages.

Thank you to my dedicated agent, Ayesha Pande, for believing in this project and also pushing me to take it to its farthest reaches. I can't imagine a better advocate for me and my work, and I feel so lucky to get to collaborate with you.

To my editor, Jean Garnett, without whom this book would have been either abandoned or unreadable: thank you a million times for working with me as I missed every deadline. Thank you for telling me frankly when things were not working. Thank you for inviting me into your home for a week to finish the thing. Thank you for your care, generosity, and immediate intimacy. I could not have written this book alone or with any other editor. And I hope one day to write like you!

Mom, Dad, and Sophia, thank you for everything. I am lucky to belong to this family, and I am honored to be shaped by your care. Grandma and Grandpa, I am in awe of all that you are and all that you have survived. Every day I hope to make you proud.

Finally, thank you, Ting, for your universe of love and care. You have supported me throughout this process in more ways than I can name, but I want to try and name some of them. Thank you for listening to and comforting me every time I fell apart. Thank you for your patience when all I could think of was work (and I promise I will never be this busy again). Your feedback made each essay richer and more honest, and so much of what I wrote is informed by your ethics and heart. It is a joy to live in your orbit with our beloveds, Sesame and Melon. Thank you for the life we share together; it is more than I ever could have dreamed.

Sources

Epigraph

Hahn, Kimiko. *Resplendent Slug*. Massapequa, NY: Ghostbird Press, 2016.

If You Flush a Goldfish

Baker, Harry. "Do Goldfish Really Have a 3-Second Memory?" Live Science, May 22, 2021. https://www.livescience.com/goldfish-memory.html.

Brunhuber, Kim. " 'Nobody Has That Much Money': One Sinking City's Fight Against Rising Sea Levels." CBC News, July 2, 2018. https://www.cbc.ca/news/world/rising-sea-levels-sfo-foster-city-1.4711621.

Clarke, Chris. "San Francisco Bay's Lost World: The Saltmarsh." KCET, June 30, 2015. https://www.kcet.org/redefine/san-francisco-bays-lost-world-the-salt marsh.

Goyal, Nikhil. "After a String of Suicides, Students in Palo Alto Are Demanding a Part in Reforming Their School's Culture." Vice, September 8, 2015.

Milliken, Randall, Laurence H. Shoup, and Beverly R. Ortiz. *Ohlone/Costanoan Indians of the San Francisco Peninsula and Their Neighbors, Yesterday and Today*. San Francisco: National Park Service; Golden Gate National Recreation Area, 2009.

Morgan, D. L., and S. J. Beatty. "Feral Goldfish (*Carassius auratus*) in Western Australia: A Case Study from the Vasse River." *Journal of the Royal Society of Western Australia* 90, no. 3 (2007): 151–56.

Morgan, David L., and Stephen J. Beatty. *Fish Fauna of the Vasse River and the Colonisation by Feral Goldfish (Carassius auratus)*. Murdoch, Western Australia: Centre for Fish and Fisheries Research, Murdoch University, 2004.

"Ohlone Land." Centers for Educational Justice & Community Engagement, University of California, Berkeley, n.d. https://cejce.berkeley.edu/ohloneland.

Rodriguez, F., E. Durán, J. P. Vargas, et al. "Performance of Goldfish Trained in Allocentric and Egocentric Maze Procedures Suggests the Presence of a Cognitive Mapping System in Fishes." *Animal Learning & Behavior* 22, no. 4 (1994): 409–20.

Rosin, Hanna. "The Silicon Valley Suicides." *The Atlantic*, December 2015.

Shirzaei, M., and R. Bürgmann. "Global Climate Change and Local Land Subsidence Exacerbate Inundation Risk to the San Francisco Bay Area." *Science Advances* 4, no. 3 (2018).

Thompson, Terry. "Fish and Game Kills Thousands of Invasive Goldfish." *Idaho State Journal,* September 26, 2020.

Tweedley, James, Stephen Beatty, Alan Lymbery, et al. "Salty Goldfish? Goldfish Can Use Wetlands as 'Bridges' to Invade New Rivers." Paper presented at the 10th Annual Wetland Management Conference, Perth, Western Australia, January 31, 2014.

Wingerter, Kenneth. "Can You Actually Keep Fish in Bowls?" PetMD, February 11, 2016. https://www.petmd.com/fish/care/evr_fi_fish-that-can-live-in-a-bowl.

Yarlagadda, Tara. "Fact-Checking the Minnesota Goldfish Mystery: Scientists Explain." Inverse, July 15, 2021. https://www.inverse.com/science/why-are-these-fish-so-big.

Yin, Steph. "In the Wild, Goldfish Turn from Pet to Pest." *New York Times,* September 22, 2016.

My Mother and the Starving Octopus

Anderson, R. C., J. B. Wood, and R. A. Byrne. "Octopus Senescence: The Beginning of the End." *Journal of Applied Animal Welfare Science* 5, no. 4 (2002): 275–83.

Bush, S. L., H. J. Hoving, C. L. Huffard, et al. "Brooding and Sperm Storage by the Deep-Sea Squid *Bathyteuthis berryi* (Cephalopoda: Decapodiformes)." *Journal of the Marine Biological Association of the United Kingdom* 92, no. 7 (2012): 1629–36.

Cosgrove, J. "No Mother Could Give More." *BCnature* 50, no. 4 (Winter 2012): 12–13.

Courage, Katherine Harmon. "Mother Octopus Sets Longest Egg-Tending Record: More than 4 Years on Baby Watch." *Scientific American,* July 30, 2014. https://www.scientificamerican.com/article/mother-octopus-sets-longest-egg-tending-record-more-than-4-years-on-baby-watch/.

Courage, Katherine Harmon. "Octopus Babies Hatch by the Thousands, Captured on Video." *Scientific American,* September 19, 2013. https://blogs.scientificamerican.com/octopus-chronicles/octopus-babies-hatch-by-the-thousands-captured-on-video-video/.

Cowles, Dave. "*Neognathophausia ingens* (Dohrn, 1870)." Rosario Beach Marine Laboratory, Walla Walla University, 2006. https://inverts.wallawalla.edu/Arthropoda/Crustacea/Malacostraca/Eumalacostraca/Peracarida/Lophogastrida/Neognathophausia_ingens.html.

Dunham, Will. "Octopus Mom Protects Her Eggs for an Astonishing 4-1/2 Years." Reuters, July 30, 2014. https://www.reuters.com/article/us-science-octopus/octopus-mom-protects-her-eggs-for-an-astonishing-4-1-2-years-idUSKBN0FZ2K920140730.

Forsythe, J. W. "*Octopus joubini* (Mollusca: Cephalopoda): A Detailed Study of

Growth Through the Full Life Cycle in a Closed Seawater System." *Journal of Zoology* 202, no. 3 (1984): 393–417.

Fulton-Bennett, Kim. "Feast and Famine on the Abyssal Plain." Monterey Bay Aquarium Research Institute, November 11, 2013. https://www.mbari.org/feast-and-famine-on-the-abyssal-plain.

Lartey, Jamiles. "Pierre Dukan, Inventor of Controversial Dukan Diet, Sued for Fraud." *The Guardian,* July 13, 2017. https://www.theguardian.com/us-news/2017/jul/13/pierr-dukan-diet-sued-fraud.

McClain, C. "An Empire Lacking Food." *American Scientist* 98, no. 6 (2010): 470.

McClain, C. R., A. P. Allen, D. P. Tittensor, et al. "Energetics of Life on the Deep Seafloor." *Proceedings of the National Academy of Sciences of the USA* 109, no. 38 (2012): 15366–71.

O'Toole, Thomas. "Octopus Surgery Has a Surprising End: Longer Life." *Washington Post,* December 1, 1977. https://www.washingtonpost.com/archive/politics/1977/12/01/octopus-surgery-has-a-surprising-end-longer-life/a8fabbce-0d76-400f-a9b4-e95b8b93094e/.

Robison, B., B. Seibel, and J. Drazen. "Deep-Sea Octopus (*Graneledone boreopacifica*) Conducts the Longest-Known Egg-Brooding Period of Any Animal." *PLoS ONE* 9, no. 7 (2014): e103437.

Seibel, B. A., B. H. Robison, and S. H. D. Haddock. "Post-Spawning Egg Care by a Squid." *Nature* 438, no. 7070 (2005): 929.

Smith, K. L., H. A. Ruhl, M. Kahru, et al. "Deep Ocean Communities Impacted by Changing Climate over 24 Y in the Abyssal Northeast Pacific Ocean." *Proceedings of the National Academy of Sciences of the USA* 110, no. 49 (2013): 19838–41.

Voight, J. R. "A Deep-Sea Octopus (*Graneledone* cf. *boreopacifica*) as a Shell-Crushing Hydrothermal Vent Predator." *Journal of Zoology* 252, no. 3 (2000): 335–41.

Wang, Z. Y., and C. W. Ragsdale. "Multiple Optic Gland Signaling Pathways Implicated in Octopus Maternal Behaviors and Death." *Journal of Experimental Biology* 221, pt. 19 (2018): jeb185751.

Yong, Ed. "Octopus Cares for Her Eggs for 53 Months, Then Dies." *National Geographic,* July 30, 2014. https://www.nationalgeographic.com/science/article/octopus-cares-for-her-eggs-for-53-months-then-dies.

My Grandmother and the Sturgeon

"Ancient Sturgeon in China's Yangtze 'Nearly Extinct.'" BBC News, September 15, 2014. https://www.bbc.com/news/world-asia-china-29201926.

Bittel, Jason. "After 140 Million Years, the Chinese Sturgeon May Soon Be Extinct." *onEarth,* National Resources Defense Council, November 20, 2018. https://www.nrdc.org/onearth/after-140-million-years-chinese-sturgeon-may-soon-be-extinct.

Cronin, John. "Sturgeon Moon." Earthdesk, August 22, 2013. https://earthdesk.org/sturgeon-moon/.

Sources

Fumei, Jiang. "Chinese Sturgeon—Aquatic Panda." *China Today,* July 10, 2018. http://www.chinatoday.com.cn/ctenglish/2018/sl/201807/t20180710 _800134891.html.

Funk, Anna. "Bad News for the Already Endangered Chinese Sturgeon." *Discover,* November 1, 2018. https://www.discovermagazine.com/planet-earth /bad-news-for-the-already-endangered-chinese-sturgeon.

Hu, J., Z. Zhang, Q. Wei, et al. "Malformations of the Endangered Chinese Sturgeon, *Acipenser sinensis,* and Its Causal Agent." *Proceedings of the National Academy of Sciences of the USA* 106, no. 23 (2009): 9339–44.

Huang, Z. "Drifting with Flow Versus Self-Migrating—How Do Young Anadromous Fish Move to the Sea?" *iScience* 19 (2019): 772–85.

Huang, Z., and L. Wang. "Yangtze Dams Increasingly Threaten the Survival of the Chinese Sturgeon." *Current Biology* 28, no. 22 (2018): 3640–47.

Kirkpatrick, Nick. "Why Did This River in China Turn Red?" *Washington Post,* July 26, 2014. https://www.washingtonpost.com/news/morning-mix/wp/2014 /07/29/why-did-this-river-in-china-turn-red/.

Knight, Tim. "Neglected Species: 'Living Fossil' Sturgeon on the Brink of Extinction." Phys.org, March 23, 2021. https://phys.org/news/2021-03-neglected -species-fossil-sturgeon-brink.html.

Lin, Xiaoyi. "10,000 Chinese Sturgeons Released into Yangtze River, the First Since Hubei Recovered from COVID-19 Epidemic." *Global Times,* April 10, 2021. https://www.globaltimes.cn/page/202104/1220690.shtml.

Long, Ben. "Extinction—by the Clock." *High Country News,* September 29, 2003.

Lovgren, Stefan. "'Living Fossil' Fish Making Last Stand in China." *National Geographic,* August 14, 2007. https://www.nationalgeographic.com/animals /article/giant-sturgeon-last-stand-china.

Peplow, Mark. "Why Has the Yangtze River Turned Red?" *News Blog, Nature,* September 11, 2012. http://blogs.nature.com/news/2012/09/why-has-the-yang tze-river-turned-red.html.

Peterson, D. L., P. Vecsei, and C. A. Jennings. "Ecology and Biology of the Lake Sturgeon: A Synthesis of Current Knowledge of a Threatened North American Acipenseridae." *Reviews in Fish Biology and Fisheries* 17, no. 1 (2006): 59–76.

"Reference: Jurassic Period." *National Geographic,* accessed May 3, 2021. https://www.nationalgeographic.com/science/article/jurassic.

"Second Sino-Japanese War: 1937–1945." *Encyclopædia Britannica,* accessed April 15, 2022. https://www.britannica.com/event/Second-Sino-Japanese-War.

"Sturgeon More Critically Endangered than Any Other Group of Species." International Union for Conservation of Nature, March 18, 2010. https://www .iucn.org/content/sturgeon-more-critically-endangered-any-other-group -species.

Sulak, Kenneth J., and Michael T. Randall. *The Gulf Sturgeon in the Suwannee River—Questions and Answers.* Report: General Information Product 72. Reston, VA: US Geological Survey, 2009.

Top China Travel, "Yichang Attractions: Chinese Sturgeon Museum." Accessed

April 15, 2022. https://www.topchinatravel.com/china-attractions/the-chinese
-sturgeon-museum.htm.

"Toxicant Is Accelerating Demise of Fossil Fish." *Science,* May 27, 2009. https://
www.science.org/content/article/toxicant-accelerating-demise-fossil-fish.

Williams, Ted. "Atlantic Sturgeon: An Ancient Fish Struggles Against the Flow."
Yale Environment 360, February 12, 2015. https://e360.yale.edu/features
/atlantic_sturgeon_an_ancient_fish_struggles_against_the_flow.

Zanon, Sibélia. "Dams Drove an Asian Dolphin Extinct. They Could Do the Same
in the Amazon." Mongabay, April 21, 2021. https://news.mongabay.com/2021
/04/dams-drove-an-asian-dolphin-extinct-they-could-do-the-same-in-the
-amazon/.

Zhuang, P., F. Zhao, T. Zhang, et al. "New Evidence May Support the Persistence
and Adaptability of the Near-Extinct Chinese Sturgeon." *Biological Conservation* 193 (2016): 66–69.

How to Draw a Sperm Whale

Bryce, Emma. "Why Was Whaling So Big in the 19th Century?" Live Science, February 22, 2020. https://www.livescience.com/why-whaling-nineteeth-century.html.

Constantino, Grace, Nick Pyenson, and Alex Boersma. "In Search of the White
Whale: A Legend, a Fossil, a Living Mammal." *Biodiversity Heritage Library
Blog,* December 9, 2015. https://blog.biodiversitylibrary.org/2015/12/in-search
-of-white-whale-legend-fossil.html.

Copeland, Jane, George Morrison, Douglas McGovern, et al. *Imprint of the Past:
Ecological History of New Bedford Harbor.* Narragansett, RI: OAO Corp.; US
EPA, 2011.

Dahlgren, T. G., A. G. Glover, A. Baco, et al. "Fauna of Whale Falls: Systematics
and Ecology of a New Polychaete (Annelida: Chrysopetalidae) from the Deep
Pacific Ocean." *Deep Sea Research Part I: Oceanographic Research Papers* 51,
no. 12 (2004): 1873–87.

Ebersole, Rene. "Why Whale Watching Is Having a Moment—in New York City."
Travel. *National Geographic,* January 5, 2021. https://www.nationalgeo
graphic.com/travel/article/why-whale-watching-is-having-a-moment-in-new
-york-city.

Ellis, Richard. "Giants of the Deep." *Los Angeles Times,* April 21, 2002.

Forde, Kaelyn, and Janet Weinstein. "Why Whales Are Returning to New York
City's Once Polluted Waters 'by the Ton.'" ABC News, August 29, 2017.
https://abcnews.go.com/US/yorks-fight-bring-back-whales/story?id
=49213546.

Fulton-Bennett, Kim. "Whale Falls—Islands of Abundance and Diversity in the
Deep Sea." Monterey Bay Aquarium Research Institute, December 20, 2002.
https://www.mbari.org/whale-falls-islands-of-abundance-and-diversity
-in-the-deep-sea/.

Goodyear, Sheena. "Orcas Now Taking Turns Floating Dead Calf in Apparent
Mourning Ritual." *As It Happens,* CBC Radio, July 31, 2018. https://www

Sources

.cbc.ca/radio/asithappens/as-it-happens-tuesday-edition-1.4768344/orcas
-now-taking-turns-floating-dead-calf-in-apparent-mourning-ritual-1.4768349.

Hanson, Brad (Leader, Marine Mammal and Seabird Ecology Team, NWFSC).
Incident report to NMFS, Office of Protected Resources, Permit and Conservation Division. *Permit 16163 Incident Report: Satellite Tag Attachment Break-
age in a Southern Resident Killer Whale and Mortality of a Previously Satellite
Tagged Southern Resident Killer Whale.* April 15, 2016.

Hume, Mark. "Orca Found Dead Five Weeks After Being Tagged." *Globe and
Mail,* April 17, 2016. https://www.theglobeandmail.com/news/british-colum
bia/orca-found-dead-five-weeks-after-being-tagged/article29656863/.

Johnson, Mark. "Experts Unravel Mystery of Blue Whale's Death." *South Coast
Today,* December 6, 1998.

Jonstonus, Joannes. *Historiae Naturalis de Quadrupetibus Libri: Cum Aeneis
Figuris.* Amsterdam: J. J. Fil. Schipper, 1657.

"Kobo." New Bedford Whaling Museum, accessed April 15, 2022. https://www
.whalingmuseum.org/learn/research-topics/whale-science/biology/skele
tons-of-the-deep/kobo-2/.

Kwong, Emily. "What Happens After a Whale Dies?" *Short Wave,* NPR, November 7, 2019.

Law, M., P. Stromberg, D. Meuten, et al. "Necropsy or Autopsy? It's All About
Communication!" *Veterinary Pathology* 49, no. 2 (2011): 271–72.

Little, Crispin T. S. "Life at the Bottom: The Prolific Afterlife of Whales." *Scientific American,* May 1, 2017. https://www.scientificamerican.com/article/life
-at-the-bottom-the-prolific-afterlife-of-whales/.

MacEacheran, Mike. "The City That Lit the World." BBC Travel, July 20, 2018.
https://www.bbc.com/travel/article/20180719-the-city-that-lit-the-world.

Mapes, Lynda V. "After 17 Days and 1,000 Miles, Mother Orca Tahlequah Drops
Dead Calf, Frolics with Pod." *Seattle Times,* August 11, 2018.

Mapes, Lynda V. "Grieving Mother Orca Falling Behind Family as She Carries
Dead Calf for a Seventh Day." *Seattle Times,* July 30, 2018.

Mapes, Lynda V. "A Mother Grieves: Orca Whale Continues to Carry Her Dead
Calf into a Second Day." *Seattle Times,* July 25, 2018.

Mapes, Lynda V. "A Mother Orca's Dead Calf and the Grief Felt Around the
World." *Seattle Times,* August 2, 2018.

Mapes, Lynda V. "Researchers Searched All Day for the Grieving Orca Mother.
Then They Found Her, Still Clinging to Her Calf." *Seattle Times,* July 31,
2018.

Mapes, Lynda V. "Southern-Resident Killer Whales Lose Newborn Calf, and
Another Youngster Is Ailing." *Seattle Times,* July 24, 2018.

Mapes, Lynda V. "UPDATE: Orca Mother Carries Dead Calf for Sixth Day as
Family Stays Close By." *Seattle Times,* July 28, 2018.

Mascarelli, Amanda. "Dead Whales Make for an Underwater Feast." *Audubon,*
November/December 2009.

"*Osedax*: Bone-Eating Worms." Monterey Bay Aquarium Research Institute, n.d.
https://www.mbari.org/bone-eating-worms/.

Pugliares, Katie R., Andrea Bogomolni, Kathleen M. Touhey, et al. *Marine Mammal Necropsy: An Introductory Guide for Stranding Responders and Field Biologists.* Technical report WHOI-2007-06. Woods Hole, MA: Woods Hole Oceanographic Institution, 2007.

Smith, C. R., A. G. Glover, T. Treude, et al. "Whale-Fall Ecosystems: Recent Insights into Ecology, Paleoecology, and Evolution." *Annual Review of Marine Science* 7, no. 1 (2015): 571–96.

Smith, C. R., H. Kukert, R. A. Wheatcroft, et al. "Vent Fauna on Whale Remains." *Nature* 341, no. 6237 (1989): 27–28.

Smith, C. R., J. Roman, and J. B. Nation. "A Metapopulation Model for Whale-Fall Specialists: The Largest Whales Are Essential to Prevent Species Extinctions." *Journal of Marine Research* 77, no. 2 (2019): 283–302.

Stefani, Giulia C. S. Good. "Losing Killer Whale L95 and Trying to Find Hope." *Expert Blog,* National Resources Defense Council, October 7, 2016. https://www.nrdc.org/experts/losing-killer-whale-l95-and-trying-find-hope.

"Whales and Hunting." New Bedford Whaling Museum, accessed April 15, 2022. https://www.whalingmuseum.org/learn/research-topics/whaling-history/whales-and-hunting/.

Yong, Ed. "The Blue Whale's Heart Beats at Extremes." *The Atlantic,* November 25, 2019. https://www.theatlantic.com/science/archive/2019/11/diving-blue-whales-heart-beats-very-very-slowly/602557/.

Yong, Ed. "The Enormous Hole That Whaling Left Behind." *The Atlantic,* November 3, 2021. https://www.theatlantic.com/science/archive/2021/11/whaling-whales-food-krill-iron/620604/.

Pure Life

Amos, Jonathan. "Humboldt Squid's Impressive Dives." BBC News, February 22, 2012. https://www.bbc.com/news/science-environment-17117200.

Ballard, R. D. "Notes on a Major Oceanographic Find." *Oceanus* 20, no. 3 (1977): 35–44.

Bittel, Jason. "New Species: Hairy-Chested Yeti Crab Found in Antarctica." *National Geographic,* June 24, 2015. https://www.nationalgeographic.com/science/article/150624-new-species-yeti-crab-antarctica-oceans.

Breusing, C., A. Biastoch, A. Drews, et al. "Biophysical and Population Genetic Models Predict the Presence of 'Phantom' Stepping Stones Connecting Mid-Atlantic Ridge Vent Ecosystems." *Current Biology* 26, no. 17 (2016): 2257–67.

Brouwers, Lucas. "Yeti Crabs Grow Bacteria on Their Hairy Claws." *Scientific American,* December 5, 2011. https://blogs.scientificamerican.com/thoughtomics/yeti-crabs-grow-bacteria-on-their-hairy-claws/.

Chave, Alan D., and Sheri N. White. "ALISS in Wonderland." *Oceanus,* Woods Hole Oceanographic Institution, December 1, 1998. https://www.whoi.edu/oceanus/feature/aliss-in-wonderland/.

"Discovering Hydrothermal Vents: 1972—The Trail Gets Hot." Woods Hole

Sources

Oceanographic Institution, n.d. https://www.whoi.edu/feature/history-hydro
thermal-vents/discovery/1972.html.

Donato, Claire. "Remembering Mark Baumer: Barefoot Walker, Poet, Climate
Activist, Friend." Literary Hub, June 10, 2021. https://lithub.com/remembering
-mark-baumer-barefoot-walker-poet-climate-activist-friend/.

"Early Clues: Red Sea 'Hot Brines.'" Woods Hole Oceanographic Institution, n.d.
https://divediscover.whoi.edu/archives/ventcd/vent_discovery/earlyclues/evi
dence_redsea.html.

Fitzpatrick, Garret. "Earth Life May Have Originated at Deep-Sea Vents." Space
.com, January 25, 2013. https://www.space.com/19439-origin-life-earth-hydro
thermal-vents.html.

Fuller, Thomas, Eli Rosenberg, and Conor Dougherty. "Fire at Warehouse Party
in Oakland Kills at Least 9, with Dozens Missing." *New York Times,* December 3, 2016.

Fulton-Bennett, Kim. "Discovery of the 'Yeti Crab.'" Monterey Bay Aquarium
Research Institute, March 2006. https://www.mbari.org/discovery-of-yeti-crab/.

"Hydrothermal Vents." Woods Hole Oceanographic Institution, n.d. https://
www.whoi.edu/know-your-ocean/ocean-topics/seafloor-below/hydrothermal
-vents/.

"Isotretinoin (Oral Route): Side Effects." Mayo Clinic, last modified February 1,
2022. https://www.mayoclinic.org/drugs-supplements/isotretinoin-oral-route
/side-effects/drg-20068178.

Kusek, Kristen M. "Deep-Sea Tubeworms Get Versatile 'Inside' Help." *Oceanus,*
Woods Hole Oceanographic Institution, January 12, 2007. https://www.whoi
.edu/oceanus/feature/deep-sea-tubeworms-get-versatile-inside-help/.

Laidre, Kristin. "Narwhal FAQ." Staff website, Polar Science Center, University of
Washington, n.d. https://staff.washington.edu/klaidre/narwhal-faq/.

Macpherson, E., W. Jones, and M. Segonzac. "A New Squat Lobster Family of
Galatheoidea (Crustacea, Decapoda, Anomura) from the Hydrothermal Vents
of the Pacific-Antarctic Ridge." *Zoosystema* 14, no. 4 (2005): 709–23.

Main, Douglas. "How the Hairy-Chested 'Hoff' Crab Evolved." Live Science,
June 18, 2013. https://www.livescience.com/37532-yeti-crab-evolution.html.

Martin, Cassie. "Life in the Hot Seat." Oceans at MIT, February 26, 2016. http://
oceans.mit.edu/news/featured-stories/life-in-the-hot-seat.html.

Meir, Jessica. "How Penguins & Seals Survive Deep Dives." Research news.
National Science Foundation, July 31, 2009. https://www.nsf.gov/discoveries
/disc_summ.jsp?cntn_id=115268.

Mullineaux, L. S., D. K. Adams, S. W. Mills, et al. "Larvae from Afar Colonize
Deep-Sea Hydrothermal Vents After a Catastrophic Eruption." *Proceedings of
the National Academy of Sciences of the USA* 107, no. 17 (2010): 7829–34.

Nevala, Amy. "On the Seafloor, a Parade of Roses." *Oceanus,* Woods Hole Oceanographic Institution, June 28, 2005. https://www.whoi.edu/oceanus/feature
/on-the-seafloor-a-parade-of-roses/.

Night Crush (@night_crush). "night crush loves, we find ourselves in another

Sources

situation where no words feel quite like the right words..." Instagram, December 3, 2016. https://www.instagram.com/p/BNk-zh2jcUU/.

Osterloff, Emily. "Hydrothermal Vents: Survival at the Ocean's Hot Springs." Natural History Museum (London), n.d. https://www.nhm.ac.uk/discover /survival-at-hydrothermal-vents.html.

Panko, Ben. "Our Oceans May Have Six Times as Many Hydrothermal Vents as Thought." *Science,* June 21, 2016. https://www.science.org/content/article /our-oceans-may-have-six-times-many-hydrothermal-vents-thought.

Pape, Allie. "San Francisco's Only Lesbian Bar, the Lexington Club, Is Closing." Eater San Francisco, October 24, 2014. https://sf.eater.com/2014/10/24/7059 907/san-franciscos-only-lesbian-bar-the-lexington-club-is-closing.

Roterman, C. N., J. T. Copley, K. T. Linse, et al. "The Biogeography of the Yeti Crabs (Kiwaidae) with Notes on the Phylogeny of the Chirostyloidea (Decapoda: Anomura)." *Proceedings of the Royal Society B: Biological Sciences* 280, no. 1764 (2013): 20130718.

Roterman, C. N., W.-K. Lee, X. Liu, et al. "A New Yeti Crab Phylogeny: Vent Origins with Indications of Regional Extinction in the East Pacific." *PLoS One* 13, no. 3 (2018): e0194696.

Rubenstein, Steve. "Oakland Fire Victim Nick Gomez-Hall: Musician and Bowler Was a 'Muse to Many.'" SFGATE, last modified December 8, 2016. https:// www.sfgate.com/news/article/Oakland-fire-victim-Nick-Gomez-Hall-musi cian-10766724.php.

Schultz, Colin. "In Defense of the Blobfish: Why the 'World's Ugliest Animal' Isn't as Ugly as You Think It Is." *Smithsonian,* September 13, 2013. https://www .smithsonianmag.com/smart-news/in-defense-of-the-blobfish-why-the -worlds-ugliest-animal-isnt-as-ugly-as-you-think-it-is-6676336/.

Scruggs, Gregory. "Seattle's Re-Bar, Marking 30 Years of Music and Weirdness, May Be Living on Borrowed Time." *Seattle Times,* February 21, 2020. https:// www.seattletimes.com/entertainment/seattle-nightlife-institution-re-bar-cele brating-30-years-of-music-and-weirdness-may-now-be-living-on-borrowed- time/.

Smith, Sable Elyse. "Ecstatic Resilience." Recess, July 2016. https://www.reces sart.org/sableelysesmith/.

Suzuki, K., K. Yoshida, H. Watanabe, et al. "Mapping the Resilience of Chemosynthetic Communities in Hydrothermal Vent Fields." *Scientific Reports* 8, no. 1 (2018): 9364.

Thatje, S., L. Marsh, C. N. Roterman, et al. "Adaptations to Hydrothermal Vent Life in *Kiwa tyleri,* a New Species of Yeti Crab from the East Scotia Ridge, Antarctica." *PLoS ONE* 10, no. 6 (2015): e0127621.

Thurber, A. R., W. J. Jones, and K. Schnabel. "Dancing for Food in the Deep Sea: Bacterial Farming by a New Species of Yeti Crab." *PLoS ONE* 6, no. 11 (2011): e26243.

"Trump's Record of Action Against Transgender People." National Center for Transgender Equality, August 20, 2020. https://transequality.org/the-discrimi nation-administration.

Weintraub, Karen. "Beaked Whales Are the Deepest Divers." *New York Times,* February 7, 2019. https://www.nytimes.com/2019/02/07/science/beaked-whales -dive.html.

Woods Hole Oceanographic Institution. "New Hydrothermal Vent Sites Found, Original Vent May Have Been Covered by Volcanic Eruption." News release. June 4, 2002. https://www.whoi.edu/press-room/news-release/new-hydrother mal-vent-sites-found-original-vent-may-have-been-covered-by-volcanic-erup tion/.

Wortham, Jenna. "The Joy of Queer Parties: 'We Breathe, We Dip, We Flex.' " *New York Times,* June 26, 2019.

Yong, Ed. "Yeti Crab Grows Its Own Food." *Nature,* December 2, 2011. https:// www.nature.com/articles/nature.2011.9537.

Zambelich, Ariel, and Alyson Hurt. "3 Hours in Orlando: Piecing Together an Attack and Its Aftermath." *The Two-Way,* NPR, June 26, 2016. https://www .npr.org/2016/06/16/482322488/orlando-shooting-what-happened-update.

Beware the Sand Striker

Anolik, Lili. "Lorena Bobbitt's American Dream." *Vanity Fair,* June 28, 2018.

Arkin, Daniel. "Lorena Bobbitt Was a Late-Night Punchline. She's Finally Getting Her Due." NBCNews.com, June 23, 2018. https://www.nbcnews.com/news /crime-courts/lorena-bobbitt-was-late-night-punchline-she-s-finally-get ting-n885721.

Baker, Katie J. M. "Here's the Powerful Letter the Stanford Victim Read to Her Attacker." BuzzFeed News, June 3, 2016. https://www.buzzfeednews.com /article/katiejmbaker/heres-the-powerful-letter-the-stanford-victim-read-to -her-ra.

Baker, Katie J. M. "Meet the Professor Who Says Sex in a Blackout Isn't Always Rape." BuzzFeed News, August 7, 2017. https://www.buzzfeednews.com/arti cle/katiejmbaker/meet-the-expert-witness-who-says-sex-in-a-blackout-isnt.

Bates, Mary. "Praying Mantis Looks Like a Flower—and Now We Know Why." *National Geographic,* December 8, 2016. https://www.nationalgeographic .com/animals/article/orchid-mantises-evolution-insects.

Black, Riley. "Giant Predatory Worms Lurked Beneath the Ancient Seafloor, Fossils Reveal." *National Geographic,* January 21, 2021. https://www.national geographic.com/science/article/giant-predatory-worms-lurked-beneath-the -ancient-seafloor-fossils-reveal.

Chozick, Amy. "You Know the Lorena Bobbitt Story. But Not All of It." *New York Times,* January 30, 2019.

"Coral Reefs," episode 3 of *Blue Planet II,* series 1. First aired February 3, 2018, on BBC One.

Cormier, Zoe. "Snapping Death Worms Can Hide Undetected for Years." BBC Earth, accessed April 17, 2022. https://www.bbcearth.com/news/snapping -death-worms-can-hide-undetected-for-years.

Crew, Bec. "This Hell Worm Dragging Prey into Its Underground Lair Is Giving

Sources

Us Anxiety." ScienceAlert, January 16, 2017. https://www.sciencealert.com
/this-hell-worm-dragging-prey-into-its-underground-lair-is-giving-us-anxiety.

Crew, Becky. "Eunice Aphroditois Is Rainbow, Terrifying." *Scientific American,*
October 22, 2012. https://blogs.scientificamerican.com/running-ponies/eunice
-aphroditois-is-rainbow-terrifying/.

"Devonian Period." *National Geographic,* May 3, 2021. https://www.national
geographic.com/science/article/devonian.

Donegan, Moira. "I Started the Media Men List. My Name Is Moira Donegan."
The Cut, January 10, 2018. https://www.thecut.com/2018/01/moira-donegan
-i-started-the-media-men-list.html.

Dunham, Will. "Unique Anatomy Helps the African Wild Dog Sustain Its Life on
the Run." Reuters, September 8, 2020. https://www.reuters.com/article/us-sci
ence-dogs/unique-anatomy-helps-the-african-wild-dog-sustain-its-life-on-the
-run-idUSKBN25Z33G.

Effron, Lauren, and Sean Dooley. "John Bobbitt Speaks Out 25 Years After Wife
Infamously Cut Off His Penis: 'I Want People to Understand...the Whole
Story.'" ABC News, January 4, 2019. https://abcnews.go.com/US/john-bob
bitt-speaks-25-years-wife-infamously-cut/story?id=60023049.

Eriksson, M. E., L. A. Parry, and D. M. Rudkin. "Earth's Oldest 'Bobbit Worm'—
Gigantism in a Devonian Eunicidan Polychaete." *Scientific Reports* 7, no. 1
(2017): 43061.

Fauchald, K., and P. A. Jumars. "The Diet of Worms: A Study of Polychaete Feeding Guilds." *Oceanography and Marine Biology Annual Review* 17 (1979):
193–284.

Finkel, Jori. "Chanel Miller's Secret Source of Strength." *New York Times,* August
5, 2020.

Fleming, P. A., D. Muller, and P. W. Bateman. "Leave It All Behind: A Taxonomic
Perspective of Autotomy in Invertebrates." *Biological Reviews of the Cambridge Philosophical Society* 82, no. 3 (2007): 481–510.

Georgiou, Aristos. "Seal Escapes Killer Whale Attack by Climbing Rocks in Tense
Video." *Newsweek,* January 4, 2021. https://www.newsweek.com/seal-killer
-whale-attack-rocks-video-1558772.

Imbler, Sabrina. "It's Not So Easy to Rename a Species with a Problematic Moniker." Atlas Obscura, May 13, 2019. https://www.atlasobscura.com/articles
/how-to-rename-a-species.

Kashino, Marisa M. "The Definitive Oral History of the Bobbitt Case, 25 Years
Later." *Washingtonian,* June 27, 2018. https://www.washingtonian.com/2018
/06/27/definitive-oral-history-of-the-bobbitt-case-25-years-later/.

Lachat, J., and D. Haag-Wackernagel. "Novel Mobbing Strategies of a Fish Population Against a Sessile Annelid Predator." *Scientific Reports* 6, no. 1 (2016):
33187.

Mah, Chris. "Who Named the Bobbit Worm (*Eunice* sp.)? And WHAT Species Is
It...Truly??" *The Echinoblog,* September 17, 2013. http://echinoblog.blogspot
.com/2013/09/who-named-bobbit-worm-eunice-sp-and.html.

Martini, Kim. "This Marine Worm Is Called the Sand Striker." Deep Sea News,

February 20, 2019. https://www.deepseanews.com/2019/02/this-marine-worm -is-called-the-sand-striker/.

Mitoh, S., and Y. Yusa. "Extreme Autotomy and Whole-Body Regeneration in Photosynthetic Sea Slugs." *Current Biology* 31, no. 5 (2021): R233–34.

Odum, Maria E. "Marine Records on Abuse May Figure in Bobbitt Trial." *Washington Post,* January 7, 1994.

Owen, James. "Mystery Solved? How Butterflies Came to Look Like Dead Leaves." *National Geographic,* December 13, 2014. https://www.nationalgeo graphic.com/animals/article/141210-butterflies-evolution-darwin-leaves -mimicry-science-animals.

Pan, Y.-Y., M. Nara, L. Löwemark, et al. "The 20-Million-Year Old Lair of an Ambush-Predatory Worm Preserved in Northeast Taiwan." *Scientific Reports* 11, no. 1 (2021): 1174.

Simon, Matt. "Absurd Creature of the Week: 10-Foot Bobbit Worm Is the Ocean's Most Disturbing Predator." *Wired,* September 6, 2013. https://www.wired .com/2013/09/absurd-creature-of-the-week-bobbit-worm/.

Simon, Roger. "Was Lorena Bobbitt's Act 'an Irresistible Impulse'?" *Baltimore Sun,* January 11, 1994.

Smith, Jonathan. "Blue Planet II: Filming Bobbit Worms in the Dark." BBC One, accessed April 17, 2022. https://www.bbc.co.uk/programmes/articles/1zzBx vhrqQRR4gpj7YG5ZjW/filming-bobbit-worms-in-the-dark.

Sokolow, Brett, and Brian Van Brunt. *Blackouts and Consent.* Online training. NCHERM Group, 2015.

Uchida, H., H. Tanase, and S. Kubota. "An Extraordinarily Large Specimen of the Polychaete Worm *Eunice aphroditois* (Pallas) (Order Eunicea) from Shirahama, Wakayama, Central Japan." *Kuroshio Biosphere* 5 (2009): 9–15.

Hybrids

Bittel, Jason. "The 'Narluga' Is a Strange Hybrid. But It's Far from Alone." *Washington Post,* June 27, 2019.

"Fishes of Kaloko-Honokohau National Historical Park: Kikakapui and Lauhau, or Butterflyfishes." *Park Species List—NPSpecies Summary Report,* December 15, 2008, accessed April 17, 2022. http://www.botany.hawaii.edu/basch /uhnpscesu/htms/kahofish/family/chaetond.htm.

Greenfield, D. W. "John E. Randall." *Copeia* 2001, no. 3 (2001): 872–77.

Helfand, Jessica. "Darwin, Expression, and the Lasting Legacy of Eugenics." *Scientific American,* August 13, 2020. https://www.scientificamerican.com/arti cle/darwin-expression-and-the-lasting-legacy-of-eugenics/.

Johnson, Akemi. "Who Gets to Be 'Hapa'?" *Code Switch,* NPR, August 8, 2016. https://www.npr.org/sections/codeswitch/2016/08/08/487821049/who -gets-to-be-hapa.

Johnson, Norman A. "Hybrid Incompatibility and Speciation." *Nature Education* 1, no. 1 (2008): 20. https://www.nature.com/scitable/topicpage/hybrid-incom patibility-and-speciation-820/.

Sources

"Lifestyle, Travel: Lizard Island." *Sydney Morning Herald,* February 8, 2004.

"Lizard Island National Park: Nature, Culture, and History." Queensland Government, Department of Environment and Science, Parks and Forests, last modified April 16, 2020. https://parks.des.qld.gov.au/parks/lizard-island/about /culture.

Montanari, Stefano R. "Causes and Consequences of Natural Hybridisation Among Coral Reef Butterflyfishes (Chaetodon: Chaetodontidae)." PhD diss., James Cook University, 2018.

NCC Staff. "On This Day: Supreme Court Rejects Anti-Interracial Marriage Laws." *Constitution Daily* (blog), National Constitution Center, June 12, 2021. https://constitutioncenter.org/blog/today-in-supreme-court-history-loving -v-virginia.

Ngai, Sianne. *Ugly Feelings.* Cambridge, MA: Harvard Univ. Press, 2007.

Nojima, Stacy. "Mixed Race Capital: Cultural Producers and Asian American Mixed Race Identity from the Late Nineteenth to Twentieth Century." PhD diss., University of Hawai'i at Mānoa, May 2018.

Online Etymology Dictionary. "Hybrid (n.)." Accessed April 17, 2022. https:// www.etymonline.com/word/hybrid.

Park, R. E. "Mentality of Racial Hybrids." *American Journal of Sociology* 36 (1931): 534–51.

Randall, J. E. "Reminiscing..." *Atoll Research Bulletin* 494, no. 3 (2001): 23–52.

Randall, J. E., G. R. Allen, and R. C. Steene. "Five Probably Hybrid Butterflyfishes of the Genus Chaetodon from the Central and Western Pacific." *Records of the Western Australian Museum* 6, no. 1 (1977): 3–26.

Rocha, L. A., A. Aleixo, G. Allen, et al. "Specimen Collection: An Essential Tool." *Science* 344, no. 6186 (2014): 814–15.

Roth, Annie. "Scientists Accidentally Bred the Fish Version of a Liger." *New York Times,* July 15, 2020.

Rowlett, Joe. "A Brief Review of Hybrid Butterflyfishes and Their Evolutionary Significance." Reefs.com, July 8, 2016. https://reefs.com/2016/07/08/brief -review-hybrid-butterflyfishes-evolutionary-significance/.

Schwartz, John. "John E. Randall, Ichthyologist Extraordinaire, Dies at 95." *New York Times,* May 29, 2020.

Turner, Ben. " 'Pizzly' Bear Hybrids Are Spreading Across the Arctic Thanks to Climate Change." Live Science, April 23, 2021. https://www.livescience.com /pizzly-bear-hybrids-created-by-climate-crisis.html.

Uyehara, Mari. "The Roots of the Atlanta Shooting Go Back to the First Law Restricting Immigration." *The Nation,* March 22, 2021. https://www.then ation.com/article/society/atlanta-shooting-history/.

Verchot, Manon. "Global Warming Spawns Hybrid Species." *Scientific American,* June 1, 2015. https://www.scientificamerican.com/article/global-warming-spawns -hybrid-species/.

"Who Was Liannaeus?: His Career and Legacy." Linnean Society of London, accessed April 17, 2022. https://www.linnean.org/learning/who-was-linnaeus /career-and-legacy.

Sources

Wilcox, Christie. "Dr. Fish." *Hakai Magazine,* March 15, 2016. https://hakaimaga zine.com/features/dr-fish/.

We Swarm

Boero, F., G. Belmonte, R. Bracale, et al. "A Salp Bloom (Tunicata, Thaliacea) Along the Apulian Coast and in the Otranto Channel Between March–May 2013." *F1000Research* 2 (2013): 181.

Chawkins, Steve. "Diablo Canyon Reactor Gets Unwelcome Guests." *Los Angeles Times,* April 26, 2012.

Colgrove, James. *Epidemic City: The Politics of Public Health in New York.* New York: Russell Sage Foundation, 2011.

Condon, R. H., W. M. Graham, C. M. Duarte, et al. "Questioning the Rise of Gelatinous Zooplankton in the World's Oceans." *BioScience* 62, no. 2 (2012): 160–69.

Dowling, Robert M. *Slumming in New York: From the Waterfront to Mythic Harlem.* Urbana: Univ. of Illinois Press, 2009.

Durkin, Colleen. "Identifying Fecal Pellets from Gelatinous Zooplankton: Pyrosomes and Salps." Plankton Ecology and Biogeochemistry, Durkin Lab at Moss Landing Marine Laboratories, San José State University, February 1, 2019. https://mlml.sjsu.edu/cdurkin/2019/02/01/identifying-fecal-pellets/.

"11 Men Arrested at Riis Park as U.S. Begins a Crackdown." *New York Times,* July 23, 1974.

Everett, J. D., M. E. Baird, and I. M. Suthers. "Three-Dimensional Structure of a Swarm of the Salp *Thalia democratica* Within a Cold-Core Eddy off Southeast Australia." *Journal of Geophysical Research: Oceans* 116, no. C12046 (2011).

Gay, Ross. "Joy Is Such a Human Madness." In *The Book of Delights: Essays.* Chapel Hill, NC: Algonquin Books, 2022.

Greenberg, Joel, Kate Madin, and Lonny Lippsett. "Salps Catch the Ocean's Tiniest Organisms." *Oceanus,* Woods Hole Oceanographic Institution, August 13, 2010. https://www.whoi.edu/oceanus/feature/salps-catch-the-oceans-tiniest -organisms/.

Henschke, N., J. D. Everett, A. J. Richardson, et al. "Rethinking the Role of Salps in the Ocean." *Trends in Ecology & Evolution* 31, no. 9 (2016): 720–33.

"The Hospital Story [from series of same title]." *The Wave,* May 17, 1956.

Kaufman, Rachel. "Mysterious Balls of Goo Are Rolling onto American Beaches." *National Geographic,* July 22, 2015. https://www.nationalgeographic.com /animals/article/150722-salp-beaches-oceans-animals-science?loggedin=true.

Klein, Joanna. "It's Better to Swim Alone, Yet Together, If You're a Salp." *New York Times,* August 4, 2017.

Law, Tara. "Ms. Colombia, Beloved Jackson Heights LGBT Figure, Found Dead." *Jackson Heights Post,* October 4, 2018.

Li, Gege. "Jellyfish Push Off a Pocket of Water Under Their Bell to Swim Faster." *New Scientist,* January 6, 2021. https://www.newscientist.com/article/2264056 -jellyfish-push-off-a-pocket-of-water-under-their-bell-to-swim-faster/.

Sources

Lorch, Donatella. "Giving Riis, the Forgotten Park, a Better Image." *New York Times,* September 7, 1991.

Madin, Kate. "Transparent Animal May Play Overlooked Role in the Ocean." *Oceanus,* Woods Hole Oceanographic Institution, June 30, 2006. https://www.whoi.edu/oceanus/feature/transparent-animal-may-play-overlooked-role-in-the-ocean/.

Madin, L. P., P. Kremer, P. H. Wiebe, et al. "Periodic Swarms of the Salp *Salpa aspera* in the Slope Water off the NE United States: Biovolume, Vertical Migration, Grazing, and Vertical Flux." *Deep Sea Research Part I: Oceanographic Research Papers* 53, no. 5 (2006): 804–19.

Maisel, Todd. "See It: 35-Foot Humpback Whale Washes Up on Rockaway's Riis Park." amNY.com, April 1, 2020. https://www.amny.com/new-york/queens/see-it-35-foot-humpback-whale-washes-up-on-rockaways-riis-park/.

"Marine Mammal Health and Stranding Response Program." NOAA Fisheries, last modified April 6, 2022. https://www.fisheries.noaa.gov/national/marine-life-distress/marine-mammal-health-and-stranding-response-program.

Mizokami, Kyle. "China's Aircraft Carriers Have a Menace: Jellyfish Swarms." *Popular Mechanics,* December 4, 2017. https://www.popularmechanics.com/military/navy-ships/a14017901/china-aircraft-carriers-jellyfish-swarms/.

"New Home for Aged Dedicated by City." *New York Times,* September 1, 1961.

New York City Dyke March website. https://www.nycdykemarch.com/.

O'Dwyer, Katie. "Meet Phronima, the Barrel-Riding Parasite That Inspired the Movie Alien." *The Conversation,* February 2, 2014. https://theconversation.com/meet-phronima-the-barrel-riding-parasite-that-inspired-the-movie-alien-22555.

Purcell, J. E., and L. P. Madin. "Diel Patterns of Migration, Feeding, and Spawning by Salps in the Subarctic Pacific." *Marine Ecology Progress Series* 73 (1991): 211–17.

Purnick, Joyce. "Koch Won't Put AIDS Patients in Queens Site." *New York Times,* September 4, 1985.

"Riis Park Beach." NYC LGBT Historic Sites Project, accessed April 17, 2022. https://www.nyclgbtsites.org/site/beach-at-jacob-riis-park/.

Rosen, Marty. "300 Stranded as City Shuts Health Center." *New York Daily News,* October 2, 1998.

Sample, Ian. "Earth May Have Been a 'Water World' 3bn Years Ago, Scientists Find." *The Guardian,* March 2, 2020. https://www.theguardian.com/science/2020/mar/02/earth-may-have-been-a-water-world-3bn-years-ago-scientists-find.

Stukel, M. R., M. Décima, K. E. Selph, et al. "Size-Specific Grazing and Competitive Interactions Between Large Salps and Protistan Grazers." *Limnology and Oceanography* 66, no. 6 (2021): 2521–34.

"Sun Bath Treatment for Tuberculous Children." *Brooklyn Daily Eagle,* July 21, 1912.

Thompson, Andrea. "Oldest Known Jellyfish Fossils Found." Live Science, October 30, 2007. https://www.livescience.com/1971-oldest-jellyfish-fossils.html.

"Urge City Use Neponsit Site." *The Wave,* January 13, 1955.

Vick, Rachel. "Beached Whale Gets Sandy Burial in Rockaway." *Queens Daily Eagle,* April 2, 2020. https://queenseagle.com/all/beached-whale-gets-sandy-burial-in-rockaway.

"The Watery World of Salps." Woods Hole Oceanographic Institution, n.d. https://www.whoi.edu/know-your-ocean/ocean-topics/polar-research/polar-life/the-watery-world-of-salps/.

Weisberger, Mindy. "1.5 Billion-Year-Old Earth Had Water Everywhere, but Not One Continent, Study Suggests." Live Science, March 2, 2020. https://www.livescience.com/waterworld-earth.html.

Williams, E. H., and L. Bunkley-Williams. "Life Cycle and Life History Strategies of Parasitic Crustacea." *Parasitic Crustacea* 3 (2019): 179–266.

Yong, Ed. "Mysterious Ocean Blobs Aren't So Mysterious." *The Atlantic,* September 26, 2016. https://www.theatlantic.com/science/archive/2016/09/these-people-can-id-the-weird-ocean-blobs-that-baffle-everyone-else/501503/.

Morphing Like a Cuttlefish

Allen, J. J., L. M. Mäthger, A. Barbosa, et al. "Cuttlefish Use Visual Cues to Control Three-Dimensional Skin Papillae for Camouflage." *Journal of Comparative Physiology A* 195, no. 6 (2009): 547–55.

Bates, Mary. "Secrets of the Flamboyant Cuttlefish's Display." *Wired,* August 27, 2014. https://www.wired.com/2014/08/secrets-of-the-flamboyant-cuttlefishs-display/.

Boal, J., N. Shashar, M. M. Grable, et al. "Behavioral Evidence for Intraspecific Signaling with Achromatic and Polarized Light by Cuttlefish (Mollusca: Cephalopoda)." *Behaviour* 141, no. 7 (2004): 837–61.

"Breeding Programs: Dwarf Cuttlefish." California Academy of Sciences, accessed April 18, 2022. https://www.calacademy.org/about-us/sustainability-in-action/breeding-programs/dwarf-cuttlefish.

Brett, C. E., and S. E. Walker. "Predators and Predation in Paleozoic Marine Environments." *Paleontological Society Papers* 8 (2002): 93–118.

Carnall, Mark. "Why Do Cephalopods Produce Ink? And What's Ink Made of, Anyway?" *The Guardian,* August 9, 2017. https://www.theguardian.com/science/2017/aug/09/why-do-cephalopods-produce-ink-and-what-on-earth-is-it-anyway.

Cartron, L., L. Dickel, N. Shashar, et al. "Maturation of Polarization and Luminance Contrast Sensitivities in Cuttlefish (*Sepia officinalis*)." *Journal of Experimental Biology* 216, pt. 11 (2013): 2039–45.

Cronin, T. W., and J. Marshall. "Patterns and Properties of Polarized Light in Air and Water." *Philosophical Transactions of the Royal Society B: Biological Sciences* 366, no. 1565 (2011): 619–26.

"Cuttlefish Males Fool Rivals by Imitating Opposite Sex." Video clip from *Nature,* season 34, episode 9, "Natural Born Hustlers: Sex, Lies & Dirty Tricks." First aired January 1, 2016, on PBS. https://www.pbs.org/wnet/nature/natural-born -hustlers-cuttlefish-males-fool-rivals-by-imitating-opposite-sex/13719/.

Ebert, J. "Cuttlefish Win Mates with Transvestite Antics." *Nature,* January 19, 2005.

Fiore, G., A. Poli, A. Di Cosmo, et al. "Dopamine in the Ink Defence System of *Sepia officinalis*: Biosynthesis, Vesicular Compartmentation in Mature Ink Gland Cells, Nitric Oxide (NO)/cGMP–Induced Depletion and Fate in Secreted Ink." *Biochemical Journal* 378, pt. 3 (2004): 785–91.

Geggel, Laura. "500 Million-Year-Old Fossil Is the Granddaddy of All Cephalopods." Live Science, March 30, 2021. https://www.livescience.com/ancient -octopus-relative-fossil.html.

Gilmore, Ryan. "Cephalopod Camouflage: Cells and Organs of the Skin." *Nature Education* 9, no. 2 (2016). https://www.nature.com/scitable/topicpage/cepha lopod-camouflage-cells-and-organs-of-the-144048968/.

Gonzalez-Bellido, P. T., A. T. Scaros, R. T. Hanlon, et al. "Neural Control of Dynamic 3-Dimensional Skin Papillae for Cuttlefish Camouflage." *iScience* 1 (2018): 24–34.

Greenwood, Veronique. "The Cuttlefish, a Master of Camouflage, Reveals a New Trick." *New York Times,* February 15, 2018.

Hanlon, R. T., and G. McManus. "Flamboyant Cuttlefish Behavior: Camouflage Tactics and Complex Colorful Reproductive Behavior Assessed During Field Studies at Lembeh Strait, Indonesia." *Journal of Experimental Marine Biology and Ecology* 529 (2020): 151397.

Hanlon, R. T., M.-J. Naud, P. W. Shaw, et al. "Transient Sexual Mimicry Leads to Fertilization." *Nature* 433, no. 7023 (2005): 212.

Jiang, M., C. Zhao, R. Yan, et al. "Continuous Inking Affects the Biological and Biochemical Responses of Cuttlefish *Sepia pharaonis*." *Frontiers in Physiology* 10 (2019): 1429.

"'Kings of Camouflage': Anatomy of a Cuttlefish." *Nova,* created March 2007. Episode first aired April 3, 2007, on PBS. https://www.pbs.org/wgbh/nova /camo/anat-nf.html.

Kröger, B., J. Vinther, and D. Fuchs. "Cephalopod Origin and Evolution: A Congruent Picture Emerging from Fossils, Development and Molecules." *BioEssays* 33, no. 8 (2011): 602–13.

Langridge, K. V., M. Broom, and D. Osorio. "Selective Signalling by Cuttlefish to Predators." *Current Biology* 17, no. 24 (2007): R1044–45.

Leibach, Julie. "Secrets of Cephalopod Camouflage." Science Friday, June 17, 2016. https://www.sciencefriday.com/articles/secrets-of-cephalopod-camouflage/.

Max-Planck-Gesellschaft. "Elucidating Cuttlefish Camouflage." News release. October 17, 2018. https://www.mpg.de/12363924/1017-hirn-080434-elucidat ing-cuttlefish-camouflage.

Max-Planck-Gesellschaft. "Passing Clouds in Cuttlefish." News release. August 1, 2014. https://www.mpg.de/8336540/colour-waves-cuttlefish.

Monks, N. "Half a Billion Years of Floating Slugs and Racing Snails: Fossil

Sources

Cephalopods FAQs." The Cephalopod Page, n.d. http://www.thecephalopod
page.org/FosCephs.php.

Nuwer, Rachel. "Biologists Are Biased Toward Penises." *Smithsonian,* May 6,
2014. https://www.smithsonianmag.com/science-nature/biologists-are-biased
-toward-penises-180951347/.

Otaka, Randy. "Capture the Iridescence of Camouflaging Cephalopod Skin." Science Friday, June 14, 2019. https://www.sciencefriday.com/educational-res
ources/capture-the-iridescence-of-camouflaging-cephalopod-skin/.

Palmer, M. E., M. R. Calvé, and S. A. Adamo. "Response of Female Cuttlefish
Sepia officinalis (Cephalopoda) to Mirrors and Conspecifics: Evidence for Signaling in Female Cuttlefish." *Animal Cognition* 9, no. 2 (2006): 151–55.

Pappas, Stephanie. "Tricky Cuttlefish Put on Gender-Bending Disguise." Live Science, July 3, 2012. https://www.livescience.com/21374-cuttlefish-gender-bend
ing-disguise.html.

Shashar, N., P. Rutledge, and T. Cronin. "Polarization Vision in Cuttlefish in a
Concealed Communication Channel?" *Journal of Experimental Biology* 199,
pt. 9 (1996): 2077–84.

St. Fleur, Nicholas. "Figuring Out When and Why Squids Lost Their Shells." *New
York Times,* March 6, 2017.

Tanner, A. R., D. Fuchs, I. E. Winkelmann, et al. "Molecular Clocks Indicate
Turnover and Diversification of Modern Coleoid Cephalopods During the
Mesozoic Marine Revolution." *Proceedings of the Royal Society B: Biological
Sciences* 284, no. 1850 (2017): 20162818.

Temple, S. E., V. Pignatelli, T. Cook, et al. "High-Resolution Polarisation Vision
in a Cuttlefish." *Current Biology* 22, no. 4 (2012): R121–22.

Thompson, Helen. "Flamboyant Cuttlefish Save Their Bright Patterns for Flirting,
Fighting, and Fleeing." *Science News,* September 1, 2020. https://www.scien
cenews.org/article/flamboyant-cuttlefish-video-mating-defense-camouflage.

Von Bubnoff, Andreas. "Playing Music Through a Squid." Science Friday, January 2, 2013. https://www.sciencefriday.com/articles/playing-music-through-a
-squid/.

Yong, Ed. "Cuttlefish Tailor Their Defences to Their Predators." *National Geographic,* May 6, 2010. https://www.nationalgeographic.com/science/article
/cuttlefish-tailor-their-defences-to-their-predators.

Yong, Ed. "Cuttlefish Woos Female and Dupes Male with Split-Personality Skin."
National Geographic, July 3, 2012. https://www.nationalgeographic.com/sci
ence/article/cuttlefish-woos-female-and-dupes-male-with-split-personality
-skin.

Zucker, I., and A. K. Beery. "Males Still Dominate Animal Studies." *Nature* 465,
no. 7299 (2010): 690.

Zych, Ariel. "Model the Texture—Changing Structures of Cuttlefish Skin: Papillae." Science Friday, June 21, 2018. https://www.sciencefriday.com/education
al-resources/model-the-shape-changing-structures-of-cuttlefish-skin-papillae/.

Sources

Us Everlasting

Azevedo, A. S., B. Grotek, A. Jacinto, et al. "The Regenerative Capacity of the Zebrafish Caudal Fin Is Not Affected by Repeated Amputations." *PLoS ONE* 6, no. 7 (2011): e22820.

Bavestrello, G., C. Sommer, and M. Sarà. "Bi-Directional Conversion in *Turritopsis nutricula* (Hydrozoa)." *Aspects of Hydrozoan Biology* 56, no. 2–3 (1992): 137–40.

Berwald, Juli. "The Immortal Jellyfish." *Discover,* November 9, 2017. https://www.discovermagazine.com/planet-earth/the-immortal-jellyfish.

Boero, Ferdinando. "Everlasting Life: The 'Immortal' Jellyfish." *The Biologist* (Royal Society of Biology) 63, no. 3 (2016): 16–19. https://thebiologist.rsb.org .uk/biologist-features/everlasting-life-the-immortal-jellyfish.

Carla', E. C., P. Pagliara, S. Piraino, et al. "Morphological and Ultrastructural Analysis of *Turritopsis nutricula* During Life Cycle Reversal." *Tissue and Cell* 35, no. 3 (2003): 213–22.

De Vito, D., S. Piraino, J. Schmich, et al. "Evidence of Reverse Development in Leptomedusae (Cnidaria, Hydrozoa): The Case of *Laodicea undulata* (Forbes and Goodsir 1851)." *Marine Biology* 149, no. 2 (2006): 339–46.

Gaskill, Melissa. "No Brain? For Jellyfish, No Problem." *Nature,* PBS.org, November 20, 2018. https://www.pbs.org/wnet/nature/blog/no-brain-for-jelly fish-no-problem/.

Gill-Peterson, Jules. *Histories of the Transgender Child.* Minneapolis: Univ. of Minnesota Press, 2018.

He, J., L. Zheng, W. Zhang, et al. "Life Cycle Reversal in *Aurelia* sp.1 (Cnidaria, Scyphozoa)." *PLoS ONE* 10, no. 12 (2015): e0145314.

Helm, R. R. "Jelly Killing Machine Tested in Korea." Deep Sea News, October 3, 2013. https://www.deepseanews.com/2013/10/jelly-killing-machine-tested-in -korea/.

Kim, D., J.-U. Shin, H. Kim, et al. "Design and Implementation of Unmanned Surface Vehicle JEROS for Jellyfish Removal." *Journal of Korea Robotics Society* 8, no. 1 (2013): 51–57.

Kramp, P. L. "Synopsis of the Medusae of the World." *Journal of the Marine Biological Association of the United Kingdom* 40 (1961): 7–382.

Kubota, S. "Repeating Rejuvenation in *Turritopsis,* an Immortal Hydrozoan (Cnidaria, Hydrozoa)." *Biogeography* 13 (2011): 101–3.

Martell, L., S. Piraino, C. Gravili, et al. "Life Cycle, Morphology, and Medusa Ontogenesis of *Turritopsis dohrnii* (Cnidaria: Hydrozoa)." *Italian Journal of Zoology* 83, no. 3 (2016): 390–99.

Matsumoto, Y., S. Piraino, and M. P. Miglietta. "Transcriptome Characterization of Reverse Development in *Turritopsis dohrnii* (Hydrozoa, Cnidaria)." *G3 Genes|Genomes|Genetics* 9, no. 12 (2019): 4127–38.

Matsumoto, Yui. "Transdifferentiation in *Turritopsis dohrnii* (Immortal Jellyfish): Model System for Regeneration, Cellular Plasticity, and Aging." Master's thesis, Texas A&M University, 2017.

Sources

Miglietta, M. P., and H. A. Lessios. "A Silent Invasion." *Biological Invasions* 11, no. 4 (2009): 825–34.

Miglietta, M. P., S. Piraino, S. Kubota, et al. "Species in the Genus *Turritopsis* (Cnidaria, Hydrozoa): A Molecular Evaluation." *Journal of Zoological Systematics and Evolutionary Research* 45, no. 1 (2006): 11–19.

Mims, Christopher. "Korea's Plan to Shred a Jellyfish Plague with Robots Could Spawn Millions More." Quartz, October 7, 2013. https://qz.com/132609 /koreas-plan-to-shred-a-jellyfish-plague-with-robots-could-spawn-millions -more/.

Osterloff, Emily. "Immortal Jellyfish: The Secret to Cheating Death." Natural History Museum, n.d. https://www.nhm.ac.uk/discover/immortal-jellyfish -secret-to-cheating-death.html.

Piraino, S., F. Boero, B. Aeschbach, et al. "Reversing the Life Cycle: Medusae Transforming into Polyps and Cell Transdifferentiation in *Turritopsis nutricula* (Cnidaria, Hydrozoa)." *Biological Bulletin* 190, no. 3 (1996): 302–12.

Rich, Nathaniel. "Can a Jellyfish Unlock the Secret of Immortality?" *New York Times Magazine*, November 28, 2012.

Schmich, J., Y. Kraus, D. De Vito, et al. "Induction of Reverse Development in Two Marine Hydrozoans." *International Journal of Developmental Biology* 51, no. 1 (2007): 45–56.

Tanaka, H. V., N. C. Y. Ng, Z. Y. Yu, et al. "A Developmentally Regulated Switch from Stem Cells to Dedifferentiation for Limb Muscle Regeneration in Newts." *Nature Communications* 7, no. 1 (2016): 11069.

Than, Ker. "'Immortal' Jellyfish Swarm World's Oceans." *National Geographic*, January 28, 2009. https://www.nationalgeographic.com/animals/article/im mortal-jellyfish-swarm-oceans-animals.

About the Author

Sabrina Imbler is a writer and science journalist living in Brooklyn. Their first chapbook, *Dyke (geology)*, was published by Black Lawrence Press. They have received fellowships and scholarships from the Asian American Writers' Workshop, Tin House, the Jack Jones Literary Arts Retreat, and Paragraph NY, and their work has been supported by the Café Royal Cultural Foundation. Their essays and reporting have appeared in various publications, including the *New York Times, The Atlantic, Catapult,* and *Sierra,* among others.